The Examination of Mines and Prospects

by C.G. Gunther

with an introduction by Kerby Jackson

Introduction

It has been over a century since C.G. Guthrie released his important publication "Examination of Prospects". First released in 1912, this work has been unavailable to the mining community since those days, with the exception of expensive original collector's copies and poorly produced digital editions.

It has often been said that *"gold is where you find it"*, but even beginning prospectors understand that their chances for finding something of value in the earth or in the streams of the Golden West are dramatically increased by going back to those places where gold and other minerals were once mined by our forerunners. Despite this, much of the contemporary information on local mining history that is currently available is mostly a result of mere local folklore and persistent rumors of major strikes, the details and facts of which, have long been distorted. Long gone are the old timers and with them, the days of first hand knowledge of the mines of the area and how they operated. Also long gone are most of their notes, their assay reports, their mine maps and personal scrapbooks, along with most of the surveys and reports that were performed for them by private and government geologists. Even published books such as this one are often retired to the local landfill or backyard burn pile by the descendents of those old timers and disappear at an alarming rate. Despite the fact that we live in the so-called "Information Age" where information is supposedly only the push of a button on a keyboard away, true insight into mining properties remains illusive and hard to come by, even to those of us who seek out this sort of information as if our lives depend upon it. Without this type of information readily available to the average independent miner, there is little hope that our metal mining industry will ever recover.

This important volume and others like it, are being presented in their entirety again, in the hope that the average prospector will no longer stumble through the overgrown hills and the tailing strewn creeks without being well informed enough to have a chance to succeed at his ventures.

Kerby Jackson
Josephine County, Oregon
January 2015

PREFACE

The purpose of this book is to present the practical side of economic geology concisely and in convenient form; established facts and the applications of accepted views are emphasized; theoretical discussions and questions of genesis are avoided. Coal, iron, and placer deposits are omitted; they are subjects of specialized study that are fully and concisely treated in other works. The reader is assumed to possess a knowledge of mineralogy, petrography, and of elementary geology.

The arrangement adopted is based on Kemp's theory of magmatic waters, Lindgren's conclusions on hydrothermal and secondary alterations, and on the theories of secondary enrichment enunciated by Emmons and Weed.

No general classification of ore-deposits is attempted, nor is any attempt made to fill by hypothesis the wide gaps, nor to explain the apparent contradictions, that rank economic geology among the most inexact of sciences. The present knowledge of the subject is too incomplete to warrant such broad generalizations.

The present situation in mining in the United States may be summed up in the statement that the demand for good properties greatly exceeds the supply. While mines of the first rank will undoubtedly be discovered from time to time, it is probably true that a great proportion of deposits having outcrops of commercial grade or of evident promise have been recognized and explored. A review of mining conditions over long periods shows that the rich discoveries belong to pioneer days, and that as time goes on the more important developments are the result of lower working costs, improved metallurgical processes, and of an increasing knowledge of economic geology.

As engineers in search of developed mines no longer expect to find properties having positive ore of greater net value than the price asked, so those in search of prospects should not expect to find proven ore-shoots awaiting their recommendation. There is usually abundant local capital for the preliminary development of a patently good prospect, and most of these are steadily worked from the time of their discovery until some apparently unfavorable development shuts off the supply of local capital. A great majority of prospects have been examined again and again, presumably by men who commanded a knowledge of sampling, the services of an assayer, and at least an elementary knowledge of geology. In order to pick a good prospect from those rejected by his predecessors, therefore, an engineer must base his hope of success upon superior geological training.

A careful search has been made of the voluminous bibliography of economic geology, and the results of this search in great part make up the present work. Individual credit is given in foot-notes wherever it may be assigned.

C. GODFREY GUNTHER.

STRATFORD, CONN., *June,* 1912.

CONTENTS

CONTENTS

CONTENTS

THE EXAMINATION OF PROSPECTS

CHAPTER I

MINING EXAMINATIONS

Mining examinations are of several kinds and the scope of the investigation depends in each case upon the purpose for which the examination is made.

Formal Examinations.—A formal examination of a developed mine is an expensive undertaking; from one to several months are allowed for the work according to the size of the property; many samples are taken; investigation is made of the geological features, including, perhaps, a topographical and geological survey of the surface; the metallurgical treatment of the ore is studied; finally, a determination of costs is demanded.

Formal examinations are made for prospective purchasers or as a basis for the consolidation of several properties, and occasionally, on the owner's behalf, to verify the work of resident engineers, or to determine the readiness of a mine for equipment and the kind and capacity of equipment it shall receive. A formal examination of a mine should not be undertaken until a preliminary examination has shown that it is justified.

It is regrettable that some engineers are in the habit of prosecuting long and exhaustive examinations of properties whose lack of merit should be, and perhaps is, apparent from the start, or of continuing their investigation long after an unfavorable result is seen to be inevitable. This wasting of a client's money is a species of trickery difficult to prove, and therefore the more contemptible.

Preliminary Examinations.—Preliminary examinations are precisely what the term implies, and are undertaken to determine the advisability of making a formal examination. There is always a reason why a mine is offered for sale, which may or may not be known from the start; frequently the chief object of the preliminary examination is to determine this reason for selling.

A preliminary examination usually includes: the cutting of a few significant samples, the number depending upon the nature and the quantity of ore claimed by the owners; a geological reconnaissance, usually involving a map; a preliminary study of the probable metallurgical treatment of the ore; and a tentative estimate of costs. By significant samples are meant samples of the claimed ore reserves, taken at regular but longer intervals than would be allowable in a formal examination, and especially samples of the lowest workings, ends of drifts, and other points where a falling off in value may be suspected.

A preliminary examination should, of course, be planned so that the results obtained may be supplemented by, and need not be repeated in the formal examination that may follow. Data that is sufficiently reliable for use in a formal report is necessary for a reliable preliminary report; the work should not be slighted because it is to be checked if it yields a satisfactory result.

Examinations for the Rescue of Badly Expended Capital.—It frequently happens that an engineer is called in only after funds are exhausted, ore reserves depleted, metallurgical processes failures, and ruin appears inevitable. This is a situation that requires special aptitude; valuable clients are at stake as well as the recovery of a part of the ill-advised outlay; furthermore, what under the circumstances may be considered a flattering salvage is almost certain to appear small to the owners after the glowing promises of the mistaken promoters.

In examinations of this kind the work should be concentrated upon the most favorable showings, a careful review of the metallurgical processes should be made, and, in fact, every loophole that promises salvage should be investigated. Where large bodies of low grade, unpayable ore have already been developed, hand picking should be tried in the hope of securing a commercial product.

The failure having perhaps been due to haphazard exploration, the geological relations of the ore should be carefully determined, and all stopes should be surveyed, as it is probable that the best ore was mined and that nothing really good was left. This data, together with assay plans of the existing ore and all geological features, should be entered upon a large-scale map, upon which relations will become clear that otherwise would not suggest themselves. A little intelligent exploration based upon such a map will in many instances yield important results.

The Examination of Prospects.—The examination of a prospect is a very different undertaking from the examination of a mine; prospects are not expected to show ore reserves as a basis for purchase, and in the last analysis the recommendation of a prospect rests on an opinion rather than demonstrable facts.

The examination of a prospect requires that all significant samples shall be taken and a thorough geological investigation be made, which need not, however, be put in formal shape unless it yields a favorable result. The question to be answered in examining a prospect is: " What chance has it to make a mine?"

The same hesitation should not be felt in recommending that a prospect be acquired under option as is justifiable before advising the purchase of a developed mine. It should be remembered that the majority of prospects have been examined many times, and that no brilliant showing of payable ore will be encountered; any fairly consistent showing of geological promise is worth considering, and merits the expenditure of a few hundreds or a few thousands of dollars in preliminary work, which, if it yields a favorable result, may be followed by more serious development.

Mining booms, even if founded on mistaken estimates, frequently lead to the discovery of valuable properties; this is due to the fact that the excitement of a boom leads to much shallow prospecting, which often exposes conditions not evident from an inspection of the immediate surface. Test pits and trenches may be considered as halfway between examination work and exploration, and while inexpensive, they not infrequently yield important results.

The Examination of Antiguas.—The presence of ancient workings should not be assumed to indicate a valuable property. Those that have had much to do with *antiguas* in Mexico have respect for the ability of the pioneers to follow, to extract, and to treat ores. These properties were worked with slave labor; a certain proportion of the force being detailed to grow food for the miners, the labor cost of mining was practically *nil;* very low grade material, therefore, could be mined, as almost any recovery from the ores was profit. Similar conditions probably obtained wherever mines were operated by the Spanish, or by the ancients.

No *antigua* should be expected to show high-grade ore; the Mexicans, for example, are expert miners, and will be found to have removed all pillars or other payable ore that was left by the pioneers in any property now sufficiently open for examination.

The first step in the examination of an *antigua* is to make a survey and to map all the workings, placing on the map all the geological data obtainable. The pioneers did not understand faults, and never drove exploratory cross-cuts; not infrequently a detailed geological study of an *antigua* results in finding important ore-bodies.

By means of samples, the grade of ore left in the mine by the pioneers can be ascertained, and this will often determine whether or not the property may be expected to yield a profit with the application of modern metallurgical processes; all ore above this minimum grade may be assumed to have been extracted.

The pioneer miners could not follow the ore very far below water level, and any *antigua*, the ore in which is primary, offers an attractive opportunity; in most cases, however, the ores mined were secondary, and probably do not continue far below the old workings before changing to low-grade primary material.

In the examination of *antiguas* the mode of bringing the ore to the surface should be carefully considered. The pioneers raised it on men's backs, and their workings—following the ore—are tortuous. It sometimes proves cheaper, where the tonnage is small, to pursue this method; the cost of a straight shaft and hoisting equipment must be apportioned against each ton of ore, and modern equipment may prove in the end more expensive than the ancient method.

If the pay shoots in an *antigua* are short and the stopes narrow, the chances are greatly against a profitable outcome for further exploration.

Price and Terms of Sale.—The final object of a mining examination being a profit for the client for whom the work is done, the price and terms asked are considerations second to none; in the United States they are too often but slightly considered, and sales of undeveloped properties are made at prices that largely discount even a decidedly favorable outcome for the proposed development.

A great majority of mining examinations lead to unfavorable reports. This is owing to the fact that the demand for good properties greatly exceeds the supply, and to the great difference between the points of view of the owners of mining properties and the examining engineers. The owners generally suffer from extreme optimism, and many engineers from excessive professional timidity, and neither is willing to meet the other halfway. Owners of prospects are usually brought to their senses after repeated unfavorable examinations, but many engineers never make a favorable report because of the risk of personal reputation. Mining is essentially a risky business, and he who declines to accept some risk will not make money for his clients.

No engineer should expect to find a mine having ore of a greater net value than the purchase price asked, unless the mine is admittedly bottomed, and has no possibilities beyond the ore already developed; this is a rare case, and most developed mines in addition to their ore reserves hold a certain promise for further tonnages of payable ore; each case must be considered on its own merits, and how much to allow for the value of prospective ore will differ with every engineer.

Mr. J. H. Curle[1] gives as his rule that 66 per cent. of the purchase price should be represented in net value of ore reserves,

[1] "The Gold Mines of the World," p. 49.

with the lower levels still looking well. On the basis of 66 per cent. of the purchase price to be represented in ore reserves, most sales are made at too high a figure, which is borne out by the fact that most mining ventures are not profitable, the balance being established by the occasional large success that repays its capital many times over.   These considerations apply chiefly to the sales of properties in an advanced stage of development and not to prospects, with which this work chiefly treats.

The Exploration of Prospects.—In the exploration of prospects it should be expected that many ventures will prove failures, as the exceptional success will more than repay a number of unsuccessful ventures. The exploration of prospects should be undertaken only when the operators have abundant capital, and the pertinacity to acquire and explore several properties successively. In such a campaign it is evident that the costs of each exploration should be kept as low as possible, and that exploration should cease as soon as the chances of success are reduced below the apparent chances when the property was acquired for exploration; persistency beyond a certain point is a failing.

In order to keep the costs low, cash payments should not be made on prospects, nor should expensive machinery be installed, as such expenditures in an enterprise of this sort must be charged against the footage of work done. It is almost always advisable at the start to follow the ore, and with as cheap temporary equipment as will answer the immediate purpose.

While it is true that the speculative chance held by a good looking prospect has a certain cash value, only in rare cases should a cash payment be made; a monthly rental, or a "salary" to the owner, during the period of exploration, frequently offers a convenient compromise.

Preliminary Search of Bibliography.—If an examination is to be made in an established district, a search of the bibliography of the district is an important preliminary to the examination. Written statements are more likely to be accurate and conservative than spoken statements, and important data are frequently obtainable. If the library is well indexed, a search of this kind is not a long undertaking.

Apex and Title.—The questions of extralateral rights and title are of prime importance in the United States. A preliminary investigation by the examining engineer usually consists in questioning residents. In the event that a sale is to be made, these points should be investigated by a lawyer. The slightest chance of legal entanglement is a sufficient basis upon which to drop a negotiation, for if a mine proves successful, the number and backing of the contestants will be multiplied.

Regularity of the Deposit.—The regularity of a deposit together with its size determines the cost of development. Of two deposits that contain a like quantity of payable ore, the regular deposit is by far the more valuable.

Unless the development of a mine is complete, which is rarely the case, the amount of ore it contains is not known until after it is worked out. A regular deposit is cheaply and rapidly developed and its value may be clearly indicated and realized through sale; a majority of irregular deposits, however, probably do not repay the cost of the development necessary to indicate their value, and must be worked by following and extracting the ore as it is found. An example of this type of property is the Butte Mine at Randsburg, California,[1] which has produced $525,000, although it is said not at any time to have had more than $5000 worth of ore exposed; many lead deposits in limestone are of this type.

Examples are numerous of deposits whose regularity permits the working of very low-grade material: the so-called "porphyry-copper" deposits are of this class, as are also deposits of the type of the Alaska Treadwell; the regularity of these deposits considered as a whole (they are locally quite irregular) permits cheap development, which in turn permits the construction of the large plants necessary to treat such low grades of ore.

Certain mines in the Catalina Mountains, Arizona, are good examples of large but irregular deposits; these properties have

[1] *Bull.* 430, U. S. G. S., p. 40.

produced large quantities of payable ore, but at no time was sufficient ore exposed to warrant the construction of a 16-mile railroad connection, or a smelting plant: as a result, the ore produced must be hauled to the railroad in wagons and shipped to a custom smelter at heavy cost, and the mines are intermittently operated.

Condition of Hanging-Wall.—A strong, firm hanging-wall is not infrequently a controlling factor in the operation of large, low-grade deposits, where the cost of timbering cannot be borne by the ore. The Granby, and other mines in the Boundary District of British Columbia, probably could not operate if it were not for the fact that their large stopes remain open without artificial support: the mines of Douglas Island, Alaska, are other examples, and the firm hanging-wall is a factor of the greatest importance in the success of the Pilares Mine, at Nacozari, Sonora, and of many other properties that contain large deposits of low-grade ore. Most of the so-called "porphyry-copper" deposits have to contend with heavy hanging-walls of soft, altered porphyry, and this disadvantage, it is thought, will become increasingly apparent as underground mining is conducted on a large scale.

Necessity for an Accurate Survey.—A survey and map are imperative to a proper understanding of any but the simplest occurrence of ore. All the geological data should be placed on the map, without which they are likely to be unintelligible. For a preliminary survey, work done with a Brunton, or other pocket transit, is of sufficient accuracy. The writer has made satisfactory maps without an instrument, using a straightedge on a large sheet of paper fastened to a smooth surface, afterward taking off the angles with a straightedge and triangle.

Refractory Ores.—That an ore is refractory is often apparent on simple inspection, as, for instance, a lead-silver ore that carries much zinc, or copper in an ore for which cyanidation would be natural treatment. A general knowledge of metallurgy is necessary to the examining engineer, and while a discussion of the details and costs of treatment are out of place in a report on a prospect, the amenability of the ore to metallurgical treatment should be borne in mind.

In the examination of a property the ore from which is to be concentrated, the probable ratio of concentration should be considered: if the valuable constituent occurs without accessory heavy metals, the ratio will, of course, be high; if the ore carries a large quantity of, for example, barren pyrite in addition to the valuable constituent, the ratio of concentration will be low, as will be also the grade of the concentrates. This condition is said to be a source of disappointment in certain large copper mines in Nevada.

Furthermore, ores whose valuable metals are present in both oxidized and sulphide form are difficult to concentrate, because of the widely differing specific gravities of their valuable minerals.

In an ore in which a part of the valuable metal is contained in soluble form, the amount so present must be considered a net loss, and the assays must be correspondingly diminished: an example of this condition that is often met in arid regions is a copper ore of concentrating grade that contains a part of its copper as chalcanthite; where such an ore exists in large deposits, leaching may offer a profitable method of treatment.

Amount of Exploration Compared with Results Attained.—The relation between the amount of work done and the quantity of ore exposed is an important one; a given quantity of ore being exposed, the result may be considered satisfactory or unsatisfactory according to whether little or much exploration was necessary to develop it

In levels from a timbered shaft it is important to note whether they were driven at regular or irregular intervals; if the latter, it is more than likely that the widest and best parts of the deposit were selected in which to drive the levels, and that they do not, therefore, indicate the average character of the deposit.

Preparing a Property for Examination.—This is a subject that the owners of mining properties would do well to study. The fact that the surface offers a complete section of the ore-deposits and of their surroundings is often lost sight of; before attempting deeper work it is well to explore the surface thoroughly with trenches and test pits. If no ore is found by this work, and no indication found of residual conditions suggesting that ore once existed at this level, then deeper exploration is not warranted.

Where the ore or gangue mineral is harder and more resistant to weathering than the enclosing rock, the outcrops are usually bold and but little surface work is necessary, but where the ore or gangue is softer than the enclosing rocks, the veins or ore-bodies do not outcrop, and surface workings are necessary to expose the deposits for examination.

The surface is always a fair criterion of the conditions in depth, if interpreted in the light of the present knowledge of impoverishment and enrichment due to surface agencies.

Development should follow the ore or vein; there is nothing more provoking than to be asked to examine a good surface showing somewhere beneath which a "cross-cut" tunnel has been driven that throws no light on the conditions at that depth, but inferentially discourages further work.

In preparing a mine for examination, the ore should be cross-cut to its full width; no engineer will allow for ore that remains "in the wall."

A Mine Dressed for Examination.—A mine that has been dressed for examination may be described as a trap for the examining engineer. The favorable features are accentuated, and the unfavorable developments are concealed,—drifts are walled up, or are allowed, or encouraged, to cave, winzes are allowed to fill with water, and workings through poor stretches of the deposit are tightly timbered to hinder examination. In stoping operations barren material is removed up to the best showings, which are left undisturbed, on the supposition that the engineer will assume that the material stoped was payable ore. Exploratory drifts are frequently stopped in good ore; it is astonishing in how many cases a man who is familiar with the local ore-shoots can stop his drift just short of running into barren ground.

Vigilance and a suspicious attitude are the engineer's safeguards against this kind of fraud. A method followed by the writer is almost sure to result in detection if trickery has been attempted. On going through the mine many questions are asked in regard to every point that suggests itself, and the answers are set down. The questions are repeated and answers noted from as many of the chief men on the ground as may be induced to answer them. After an interval the questions are repeated, but not in the same order, and the answers are again noted. Before leaving the property a third inquisition will yield the desired result. If the local representatives have been telling the truth, their answers will check up, but it is a very exceptional liar that can stick to a fabric of falsehoods three separate times with long intervals between.

Past Production.—A study of smelter or mill returns from past ore shipments in connection with the amount of exploration done is often instructive in the examination of irregular deposits. The average value of past ore shipments, of course, is no criterion of the grade of ore left in the mine, as the best ore is invariably extracted first.

Low-Grade Ores.—A slightly explored property carrying low-grade ore that may be expected just to pay its way should be explored, as higher-grade ore may be found. Any large deposit of even very low-grade material should be given at least a preliminary examination, as the constant improvement in metallurgical processes is steadily rendering payable ores of lower and lower grade. The waste of 15 years ago is the ore of to-day, and the same advance may be reasonably expected in the future.

Large Versus Small Properties.—In general, a large body of low-grade ore is more likely to persist than a smaller body of high-grade ore. Small mines, unless containing shipping ore and requiring little equipment (and these are rarely purchasable at a reasonable price), are usually not profitable. The cost of all equipment must in the final analysis be charged against the tonnage of ore extracted, and one mine may yield a handsome profit where two mines each containing one-half the tonnage of the same grade may both net losses. Furthermore, a small property is in a poor strategic position to secure good smelting or freight rates.

Sampling.—A discussion of sampling, although fully treated elsewhere,[1] cannot logically be omitted; inasmuch as proper sampling is often neglected, the subject will bear repetition.[2] Sampling is expensive work if properly carried out, and no other kind of sampling is of any value.

The ideal sample is a uniform groove, or channel, across the full width of the ore, and no more; how closely this may be approached in practice will depend upon the material sampled and upon the time and care given to the work.

The Equipment for Sampling.—A hammer and moil are preferable to any other tools in cutting samples; a prospector's pick will do good work in soft, uniform ground, but in harder material, even if the po'nt and hammer-end are alternately used, is likely to have a selective effect, and samples taken with a pick are not above suspicion.

To catch the sample a cloth is best, spread out so as to catch all chips. If the ground is loose and masses are likely to fall, the sample is best caught in a box, which is also used where fine, rich material is likely to sift out of cracks and vugs and so find its way into the sample.

To break down samples a crusher is convenient, but two large, tough stones of barren rock such as may be found in any creek bed, one to lay the ore upon and the other to pound with, yield the maximum result for coarse breaking; unless the ore is very hard, the abrasion of the stones is negligible.

To cut down samples, rolling and quartering on flexible oil-cloth is a good method, but where many samples are to be taken,

[1] "Sampling and Estimation of Ore in a Mine," T. A. Rickard and others.

[2] Credit cannot be given to all those that have written on this subject for the many practical points that come up as obvious solutions of the minor problems of mining examinations.

a Jones sampler with four pans, or a riffle, is quicker and more certain to do accurate work.

A simple apparatus to permit the cutting of samples in an untimbered shaft has been used by the writer with great saving of time and expense over erecting platforms. This consists of a short seat, about 14 in. long by 6 in. wide, to which is rigidly fastened a pole in length about one and one-half times the width of the shaft to be sampled. This seat is fastened to the end of the windlass rope, and, straddling the rope, the sampler is lowered to the point where a sample is to be taken, meanwhile holding the pole parallel to the rope so as to permit the descent. Arriving at the point where the sample is to be taken, the pole is allowed to fall against the opposite wall, the end of the pole catching in an inequality of the rock, the seat is hoisted a few inches, and the sampler, with feet braced against the face sampled, is held firmly and safely in position, with both hands free to work. The sample may be caught in a bag

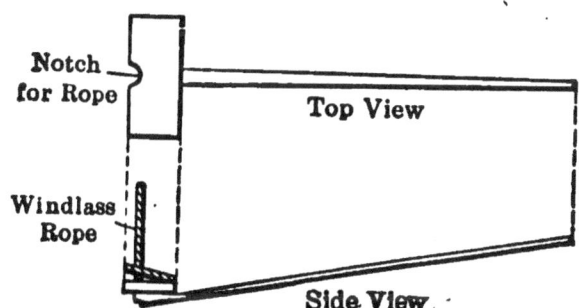

FIG. 1.—Sketch showing sampling seat for use in untimbered shafts; the pole should be bolted firmly to the seat.

held between the feet, or a canvas receptacle may be rigged in front of the sampler, or the sample may be allowed to fall on a canvas spread over the bottom of the shaft, which should be protected from the impact of the falling ore by a few boards; the latter is usually not a safe method unless the men at the windlass are closely watched, as salting by them would be an easy matter.

Marking the Samples.—Durable tags carrying the number of the sample should be inserted in each sack; some engineers use metal tags, some wood, but the usual practice is to use tough paper rolled up tightly to prevent abrasion. The following sample tag is excellent, and should be made up in books containing 50 sheets and numbered before going into the mine.[1] The

[1] R. C. Gemmel, "Sampling and Estimation of Ore in a Mine," p. 185.

lower part of the tag is torn off, and after being rolled up, is inserted in the sack with the sample. Tearing the detached slip in half gives two numbers for the duplicates when the sample is

> Date...........
> Sample taken............................
> At point...................................
> From......................................
> Across....................................
> For........... Ft............. In......
> Measurement: At right angles to dip, vertical, horizontal.
> To..
> Dip................. Strike...........
> No.................................
> ————————— (perforation here) —————————
> No..................
>   No.............

cut down. On the backs of these slips the writer is accustomed to make a sketch in section of the place sampled, showing the shape of the drift and the sample cut by a dotted line; the geological features may also be indicated with colored pencils; and any remarks noted. The data is thus kept in convenient form for future reference and is invaluable in case of dispute; it frequently happens that an unfavorable report is questioned or disputed, and in any controversy regarding samples these books so kept will put the adversary to rout on sight.

The writer places the sample number on the outside of the sacks to permit ready identification without having to pour out the sample in search of the tag. This practice is objected to by some engineers on the ground that it permits an outsider to locate the samples with equal ease. If an outsider gets close enough to the samples, and the leisure in which to inspect the numbers, he is close enough to tamper with them, and this objection is not, therefore, considered as having weight.

Preparation to Sample.—Samples should be accurately referred to some permanent object, such as a cross-cut, winze, or survey station. The intervals between samples should be measured along the center of the drift, as they otherwise will differ widely according as they are measured on one side of the drift or the other, and will therefore fail to plot correctly on the map.

It is poor judgment to mark the points at which the samples are to be taken in advance of the actual sampling; this amounts to an advertisement that a sample is to be taken along a certain line, and permits the evilly disposed to assist nature in the distribution of values.

A face that is to be sampled should be thoroughly cleaned. If the ground is soft, a strip a few inches wider than the sample cut should be cleaned off with a pick; if the ore is hard, a brush or broom should be used either dry or with water. In driving any working, the fine material, often the richest, is powdered and thrown against the roof and walls, where a portion of it adheres; it is, therefore, of the greatest importance that the face to be sampled should be thoroughly cleaned. Irregular projections and loose pieces should be knocked off, to give, in so far as possible, a flat surface from which to cut the sample.

The face to be sampled should be examined carefully for soluble salts. In copper mines in dry climates an efflorescence of chalcanthite and other salts is usual, and the sampling of old workings is attended with considerable risk of salting from this cause. Samples containing these efflorescences, even after boiling in water, show blue crystals under the microscope. These efflorescences are due to the evaporation of migrating solutions on the walls of the workings, and represent an enrichment not found throughout the mass of rock. The writer had occasion to resample a mine where there was much efflorescence. His samples were boiled in water with a little caustic soda and averaged about 3/10 per cent. copper; the sampling thus discredited averaged in excess of 2 per cent.

Where it is necessary to take samples from the floors of drifts it is best to cut large samples and to wash from them and disregard all fine material; fine particles of heavy minerals work into cracks in the floor and give deceptively high results. This method may give results somewhat too low, but is not so liable to serious error as would result from the inclusion of the fine material.

Samples should always be taken as nearly as possible at right angles to the lines of distribution of the minerals through the ore.

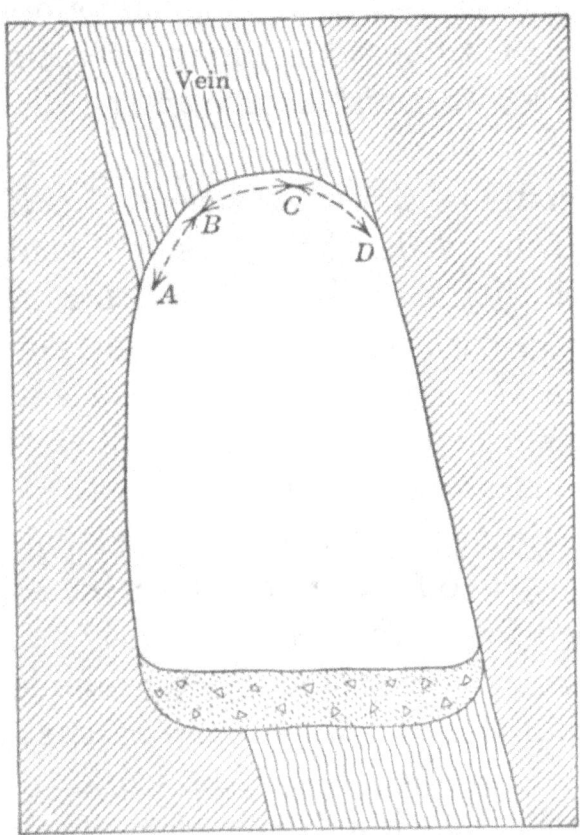

Fig. 2.—Cross section showing irregular exposure of ore in the roof of a drift; the quantity of sample taken per foot should be less along *A–B* and *C–D* than along *B–C*, where the roof is approximately at right angles to the vein.

Placing the Samples.—The interval between samples depends upon the regularity with which the valuable minerals are distributed. One extreme might be considered an absolute'y uniform mass, of which one sample would suffice, and the other extreme, a segregation of all the valuable mineral into a single mass; it is therefore apparent that the proper interval between samples will differ with each exposure sampled.

In general, a 20-ft. interval will suffice for a large ore-shoot of uniform grade, a 10-ft. interval in average cases where many samples are to be taken from the same ore-shoot, and 5-ft. or lesser intervals where the ore is spotty. It is usually advisable to start with 20-ft. intervals in the examination of a property where no data are available, and resample at 10-ft. intervals, and perhaps again at 5-ft. intervals, where the results from the first series indicate that such a course is advisable.

In sampling a wide vein or deposit it is best to divide the width into sections and to sample them separately, in order to determine the distribution of the values. These widths may be taken over even multiples of the total width if the deposit presents a uniform appearance. If the vein or deposit presents a variegated or banded appearance, however, the several bands or zones should be sampled separately.

Where a section of ore is irregularly exposed, as is commonly the case with a vein in the roof of a drift, the sample must be cut deeper over the part that is at right angles to the vein than where the face is slanting, in order not to get an undue proportion from the slanting exposure.

The Size of Samples to be Taken.—The size of a sample should be limited to the least amount that will yield a true average of the exposure sampled. A few large samples are of little value as compared with many smaller samples, if the latter be well taken. Car-load shipments, ton-samples, and shooting down large samples are obsolete methods, as a bunch of rich ore is capable of salting the whole sample, and any sample that is too large to be sealed in a sack and properly protected against salting is a source of danger. Small mill runs are not satisfactory on any but very high-grade ores, as the clean-up will depend largely upon whether the plates are scraped clean or whether they are allowed to absorb amalgam.

The more uniform the ore the smaller may be the samples; where the ore is spotty, the samples should be large, as is also the case where the ore is loose and breaks irregularly, or is alternately hard and soft.

In cutting a sample the rich spots should be avoided if they are few in number; if there are many rich spots, the groove should include everything along its line.

It is a very difficult matter to sample correctly a spotty ore. In the case of an ore that consists of barren or nearly barren quartz carrying free gold, the average cannot well be determined by sampling; a majority of the samples of such an ore will be very low, or blanks, and a few will be very high; the average obtained will be more a matter of luck than a basis for an accurate estimate; a large mill run is the solution of this problem.

Salting.—The best safeguard against salting is to decide upon a plan for safeguarding samples and never to make any exception in carrying it out, whatever the circumstances; if an engineer trusts his judgment as to whether he is in safe company, his judgment is almost certain to be at fault at some time during his career, but if he always maintains the same vigilance he will never be salted. Salting is the result of carelessness, and is inexcusable.

The inclusion of waste samples is generally recommended as a safeguard, but like much good advice, is rarely carried out.

The writer insists that no one except assistants whose integrity is known to him shall approach a sample until it is placed in a new, clean sack, and sealed, with the top turned over and tied down to prevent the working in of fine particles at the mouth of the sack. As a check upon a series of samples so taken, it is well to save a portion of the fines from rejected quarterings and to pan them; if a black greasy scum appears, it is evidence of tellurides; if a string of colors appears, its origin should be ascertained.

After being sacked the samples should be locked in a mail sack, preferably made of leather, or in a trunk. If the sacks used are clean, a syringe cannot be used without detection through the stain left on the inside of the sack.

It is much better to offend the vendors by the precautions taken than to run any chances of being salted; those with honest intent seldom take offense at such precautions.

The Shipment of Samples.—Before shipping samples it is best if possible to grind them to a degree of fineness such that the assayer is sure to take a representative portion for assay; this is hard work, but it should not be left to the assistant of the assayer employed, who may not give proper care to this important work.

In shipping samples it is well to direct the assayer to reserve the pulp in case an umpire assay should be required, which will indicate to him that he is assaying against another man and will so induce accuracy. A duplicate set of samples should always be kept by the engineer. With low-grade ores it is best to instruct the assayer to make crucible assays on two assay-ton lots. Samples should be packed for shipment in boxes in preference to sacks.

It should be unnecessary to state that local assayers should be viewed with suspicion, and that the man who is to run the samples should be as thoroughly known to the engineer as any assistant whom he may entrust with the sampling. In large examinations an assayer is usually included in the staff.

Resampling.—It is usually advisable to resample personally a certain proportion of the cuts as a check upon assistants, as well as to check high assays.

High Assays.—The treatment accorded abnormally high assays will vary with each property examined and with every engineer. The usual procedure is to reassay the sample one or more times, to determine if the high result is due to a rich speck in the pulp taken for assay. If the sample as a whole is found to be high, the cut should be resampled. If this result checks the first, some engineers recommend resampling halfway between the high sample and the adjacent samples, and using the average of these results in place of the high result; others advocate the omission of the high assay, using in its place the average of the other assays from the same exposure.

The most reasonable basis upon which to consider a high assay is in the light that it is due to the average grade of ore plus an extra amount of the valuable mineral, and to substitute for it the average of the higher samples from the same exposure the results of which have been accepted.

The Calculation of Results.—The foot-ounce method is the one generally adopted in the calculation of ore reserves. Several elaborate and complicated methods have been put forward by various engineers, but it seems probable that the results of calculation by the foot-ounce method are as accurate as the results of the samples themselves. In this method the length of each sample is multiplied by its assay value; the products from all the samples in the block under consideration are added, and this total divided by the sum of the lengths, the quotient being the average value.

All calculations should be made in dollars for gold, in ounces for silver, and in percentages for other metals, and these values should not be translated into dollars per ton until the final result for a given block is obtained, when the market price used for the various metals should be set down also.

In calculating tonnages the specific gravity of the ore should be carefully determined; it is not unusual for engineers to assume an average specific gravity for the ore, a procedure that is likely to lead to serious error; this is apparent if it is remembered that the percentage error in the ore reserves is directly proportional to the percentage error in the guess at the specific gravity. With ores composed largely of heavy sulphides, the specific gravity may be determined by estimation, which requires that the percentage of iron be determined in addition to the other base metals; this method will yield accurate results and should be used where the specific gravity of the ore varies greatly, the calculation being applied to the individual sample results. The best method in most cases is to determine the specific gravity directly, weighing in air and afterward in water several batches of material from different parts of the ore-shoot under investigation. The method of packing a box with ore and determining the weight of a known volume of broken ore, and then introducing a factor to represent the relative volumes of broken ore and ore in place, is open to objection; the factor introduced is a guess, and the result depends upon the ratio of voids to ore and therefore upon the tightness with which the ore is packed into the box.

The determination of the specific gravity of a porous ore is a difficult matter; it may best be accomplished by weighing in air and determining the volume of the pieces weighed by overflow of a vessel full of water. If this is done rapidly the result is correct, but does not take into account vugs or open spaces, which must be allowed for.

Stoping Width.—The relation between the width of a vein and the stoping width necessary to extract the ore, should receive careful attention in the examination of narrow veins. The minimum stoping width for machine drills is usually from 4 to 5 ft.; the new air-hammer drills require less. For hand drilling, 30 in., or sometimes even less, is required.

The stoping width and the amount of waste broken with the ore varies with the relative hardness of the ore and wall rocks, and also according to the firmness of the hanging wall, which, if loose, will contribute waste by caving. In some veins the waste may be shot down first and the ore broken down clean afterward.

In the consideration of narrow veins the average amount and grade of ore must be dertemined as it stands, and the values, if any, that are carried by the wall rock; then, upon the assumption of a stoping width, the average value of the broken ore may be calculated. This average should be further corrected by the amount and value of the waste that may economically be sorted out, the final result being the grade that may be expected for the ore to be hoisted or shipped. These figures, in the case of an operating mine, may be checked by records of past production, where the volumes of material stoped can be determined, or where the amount and grade of the waste sorted out is known.

Hand Picking.—The results indicated by sampling are almost always diminished by slabs of wall rock that unavoidably become mixed with the ore. In deposits that carry heavy sulphides unevenly disseminated through the gangue, and in ores through which the values are irregularly distributed in such a manner that the richer portions are distinguishable to the unaided eye, the question of hand picking is of prime importance.

Hand picking is a very effective process, where properly arranged for and carried out on clean ore from which the fine material has been screened. Not only is the waste sorted out in hand picking, but clean mineral is saved as a high-grade product. The process yields two clean products, finished ore and clean waste, in one operation, and thereby saves not only mill capacity, and the cost of treating the waste, but also greatly reduces the metallurgical losses that the clean mineral would otherwise suffer. An inexpensive installation to permit efficient sorting will often result in the recovery of a payable product from an ore that could not otherwise be considered an asset.

Metallurgical Losses.—The question of metallurgical losses is as important as that of the average grade of the ore; in all cases it should receive careful attention, and in formal examinations an exhaustive series of tests may be necessary. The mineralogical character of the ore is the criterion in the field work, unless laboratory tests can be made. Before a plant is installed or a developed property purchased, however, large scale tests are advisable.

The Cost of Mining and Treatment.—The cost of mining and treatment is a factor as important as grade of the ore, and is probably more subject to the personal equation than any other branch of an examining engineer's work: individual judgment based on experience is the final guide in an estimate of costs. This subject, which applies rather to the examination of developed mines than to prospects, is most thoroughly gone into in **Mr. J. R. Finlay's** book on "The Cost of Mining."

The results of past operations must be considered in the light of future probabilities, and it should be borne in mind that mining costs and selling prices of metals vary greatly with time, and, in some cases, with the seasons.

The larger the property and the more complete its development the easier becomes the problem of estimating costs. Those properties that most nearly approach the character of a manufacturing enterprise are the easiest with which the engineer has to deal.

In the examination of an isolated prospect having no ore reserves, a detailed discussion of probable costs is out of place as the costs will vary according to the tonnage developed. A tentative estimate based on experience is the best that may be offered in such cases.

The regularity of a deposit and its absolute size are important factors in the cost of mining; the angle of inclination of the deposit with the horizon is important, also, as determining whether the ore when broken in the stopes will run or whether it will have to be shoveled.

All the other attributes of a prospect must be considered in the light of its situation with respect to transportation; excessive distances, or a rugged topography without roads or trails, ma render valueless a property that would be valuable if better situated.

Water is a necessity for camp use and for metallurgical plants, and if there is no visible source of water near a prospect, the question at once becomes grave. Some gold mines in the Altar District of Sonora, and in many other desert regions, are commercially impossible on account of scanty water supply. Water in old workings should be investigated, and its source determined— whether it is due to seepage, or whether it comes from an underground channel. The amount of flow of water is, of course, of vital importance, and the question as to whether the flow comes from an underground reservoir which may ultimately be drained, or from a regularly flowing channel, is important. Examples are numerous where large flows of water have rendered impossible the mining of otherwise valuable ore-bodies.

Climate and altitude are not usually controlling factors in the United States, but become such in the high latitudes, or in the fever belts in the tropics. The high altitude has been a great detriment to the development of certain districts in southwestern Colorado.

Wages and supply of labor are factors for careful consideration; fuel and motive power also are factors of prime importance.

It is often advanced that a company, through large expenditure for equipment, may greatly better the working costs of a property that has been operating on a small scale; this is rarely true with small high-grade mines, where the small owner can secure working costs comparing favorably with anything that a large company can accomplish. This is the reason that small high-grade mines are usually poor purchases; their owners demand all that they are, or may become, worth. Good examples of this condition are offered by many mines in Mexico that produce ore as cheaply with primitive methods as could a large company with expensive installations.

The cost of equipment must finally be charged off against the tonnage mined, and for this reason large low-grade properties are likely to offer the best opportunities to the average investor.

In determining the probable cost of mining, the development cost must not be forgotten; the footage of shafts, drifts, raises, and so forth must be considered in connection with the amount of ore developed, and the probable cost of future development must be included in the estimate of the mining costs.

Finally, the purchase price is a charge of which each ton must bear its share. These factors are often forgotten in America, where mining risks are too often accepted in a gambling spirit.

The Estimation of Ore Reserves.—Most engineers will agree fairly well in regard to the quantities of developed ore and of probable ore in any mine, but great differences are to be expected in estimates of possible ore, which, of necessity, are forced predictions of the future based upon insufficient data. A reasonable method by which to evade such predictions is suggested by Mr. J. H. Curle;[1] it is to purchase mines on the basis of the ore reserves plus a royalty for all the ore afterward discovered.

The starting-point in a consideration of the probable depth to which an ore-shoot will persist is the determination whether the ore is of primary or of secondary origin; if primary, there is no genetic reason why the shoots should not continue to great depths; if secondary, little or no tonnage may be allowed below the lowest workings, unless the mineralogical character of the ore is such that it seems reasonably certain that the workings are still in the upper part of the zone of enrichment. Secondary ore-shoots, in general, are greater in horizontal than in vertical extent; primary ore-shoots, on the contrary, are commonly greater in vertical than in horizontal extent; the relation between the horizontal and vertical dimensions of exposed ore-shoots should be considered in making estimates of the depth to which they probably continue.

If the ore exposed is the result of geological conditions unlikely to be duplicated in depth, the end of the mine is probably in sight.

It is certainly true that the tendency of primary as well as of secondary ore-shoots is to pinch out or to become low in grade

[1] "The Gold Mines of the World," p. 46.

with increasing depth; many primary ores, however, probably continue to depths below the limit of mining.

A factor of safety for the engineer's personal reputation has no place in a mining report, except in the margin of profit that he deems necessary to make the enterprise attractive.

CHAPTER II

STRUCTURAL GEOLOGY

Ore-deposits are commonly divided into two classes, syngenetic and epigenetic, according to whether the ore was deposited together with the enclosing rock or was introduced after its deposition or solidification. Epigenetic deposits, which are by far the more important class, owe their formation to the channels that permitted the ingress of their metals, and both classes are subject to great modification by post-mineral changes in the containing rocks. A discussion of the structural features of rocks, therefore, necessarily precedes any consideration of ore-deposits or processes of ore deposition.

Stratification.—The arrangement of sediments in parallel and approximately horizontal layers is called stratification, and is an original property of sedimentary rocks. Successive strata of the same sedimentary bed commonly differ from each other in some minor characteristic, such as texture or color, which results in a bedded or stratified appearance.

Cleavage.—The subjection of a rock to pressure develops within it parallel planes of weakness, called cleavage planes, along which the rock breaks into relatively regular slabs or blocks; the direction of these cleavage planes, which bear no relation to stratification, is determined by the direction of the pressure that produced them. Where there are more than one set of cleavage planes, one of them is likely to be more prominent than the others. In sedimentary rocks, the cleavage may coincide with the stratification, but more commonly cuts across it. Cleavage has been defined[1] as the "capacity present in some rocks to break in some directions more easily than in others." Cleavage, therefore, does not imply the existence of subdivisions, but rather the tendency to subdivide along certain planes.

[1] Van Hise.

Fissility.—Fissility is a "structure in rocks by virtue of which they are already separated into parallel laminæ."[1] Fissility may be regarded as a development of the property of cleavage, and is commonly expressed along closely set parallel planes.

Schistose Structure.—The long-continued stresses of regional metamorphism with accompanying recrystallization and rock flowage develop a banded structure in both sedimentary and igneous rocks; the various minerals contained by the rocks are arranged with their longer axes parallel and form planes of weakness to fracture that differ from cleavage planes and that are independent of any original stratification. A rock that has suffered this change is said to have a schistose structure. Schists commonly exhibit cleavage that has no relation to their schistosity, and occasionally, also, traces of the original stratification are visible.

Gneissic Structure.—A coarsely schistose structure is known as a gneissic structure. In gneisses the individual bands are more prominently developed than in schists, and the ease of fracture along these bands is relatively less.

Joints.—Planes of division through rocks that are the result of stresses insufficient to produce more than microscopic movement are called joints. Fragments of broken rock, both sedimentary and igneous, are commonly bounded in part by plane surfaces, which are joint planes. Joints are of several kinds, according to the nature of the stresses that produced them; they vary in expression from cleavage, or incipient jointing, to the well-developed planes that bound the prismatic columns of certain basalts. Joint planes frequently preserve their general directions over long distances, and the angles between different joint systems are likely to be constant throughout large rock masses.

Where the joint planes are closely spaced and where the load of overlying rocks is light, they afford channels for the circulation of solutions and frequently become mineralized.

During the contraction that results from cooling and solidification, igneous rocks separate into masses of roughly polygonal

[1] Van Hise.

section bounded by tension or contraction joints. Such joints are best developed in certain flows, which upon solidification divide into regular prismatic columns.

Sedimentary rocks upon drying not infrequently suffer contraction, which finds expression in similar, but usually less well-defined, planes of division. Fractures due to contraction are the result of internal strains in a rock mass, and do not pass outside of it into other rocks for any important distance.

FIG. 3.—Slates on Elk Creek, Idaho, showing joint planes at nearly right angles to the bedding. *After Ransome.*

The outer members of folded strata undergo tensile stress during folding, which occasionally results in the formation of tension joints that follow a radial arrangement inward from the arc of the fold.

Sheeting.—Parallel planes of fracture, developed by compressive stresses, of relatively great continuity as compared with joints, but of only incipient displacement, are called sheeting planes. Closely set and well-developed sheeting planes often afford

FIG. 4.—Devil's Tower, Wyoming, showing columnar jointing of igneous rock. *After Darton.*

channels for the circulation of solutions, and, when mineralized, form sheeted lodes. Not infrequently systems of sheeting planes occur in pairs, parallel in strike, but intersecting in dip, such

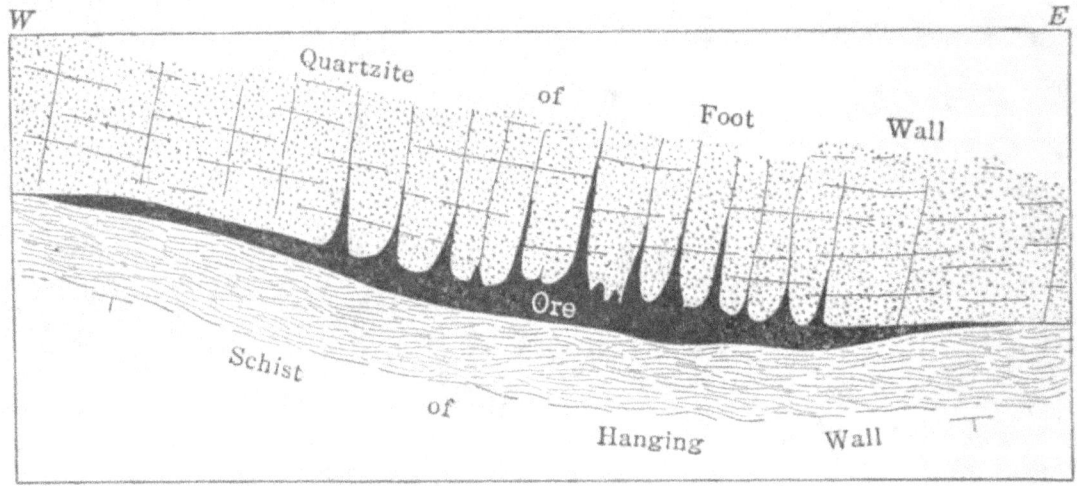

FIG. 5.—Ideal section of the Ferris-Haggerty ore-body, Encampment, Wyoming, showing convexity and the mineralization of tension joints. *After Spencer.*

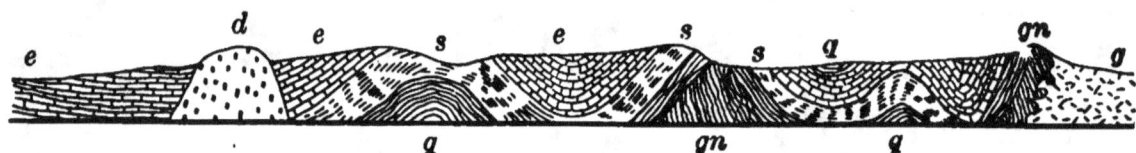

FIG. 6.—Section of folded strata near Negaunee, Michigan. *gn,* Gneiss; *g,* granite; *q,* quartzite; *s,* clay slate; *e,* iron-bearing Negaunee strata; *d,* diorite and diabase. *After Van Hise and Bayley.*

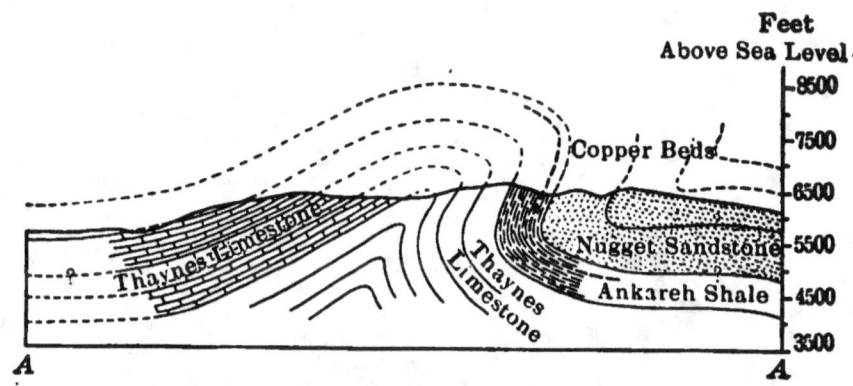

FIG. 7.—Partly eroded anticline, Montpelier, Idaho. *After Gale.*

fracturing being the typical result of compressive or torsional strains; these interdependent systems of sheeting planes are known as conjugate systems.

Folds.—A fold is a bend in a rock mass caused by compressive stress of insufficient intensity to produce a fault. A fold is called a syncline if its bend is concave above, and an anticline if its bend is concave below. A dome is an anticline whose length is zero,

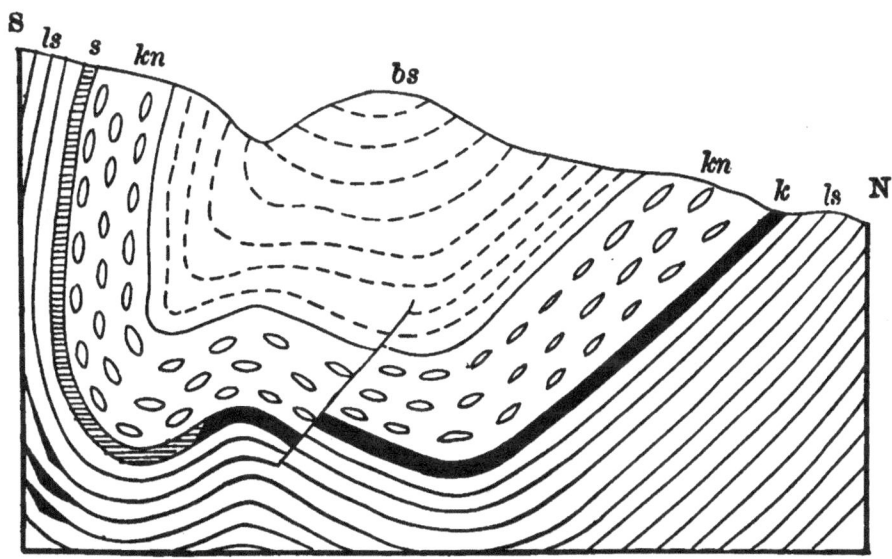

Fig. 8.—Section through the ore-deposit at Meggen, Germany, showing a synclinal trough. *k*, Pyrite; *s*, barite; *ls* and *bs*, slates; *kn*, limestone. *After Hundt.*

and is best described by the usual meaning of its name. A basin is the synclinal equivalent of a dome. Folds are commonly persistent in strike, and domes or basins are of relatively rare occurrence. A monocline may be described as a half-fold, as

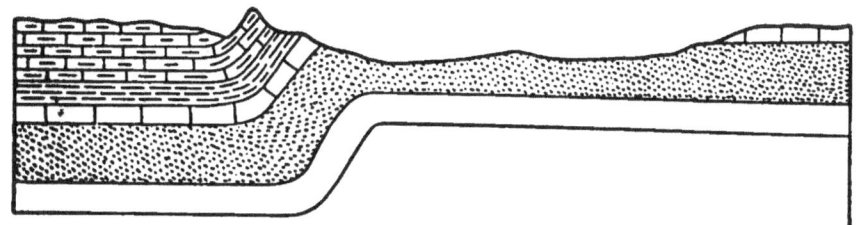

Fig. 9.—Monocline, near Gallup, New Mexico. *After Howell.*

where strata assume a terrace-like position, being parallel at different elevations either side of an inclined connecting part. The rock forming the convex or outer portion of a fold is, like the lower portion of a beam subjected to load, under tensile stress, which frequently results in a series of fractures that may afford channels for the circulation of solutions and so become *loci* of ore deposition.

The Development of Faults from Folds.—Where a fold is formed by compressive stress beyond the capacity of the rock to withstand, a fault is developed along the axis, or the plane bisecting the angle between the component limbs of the fold. Folds not infrequently pass into faults along their strike, and in a region that has been subjected to

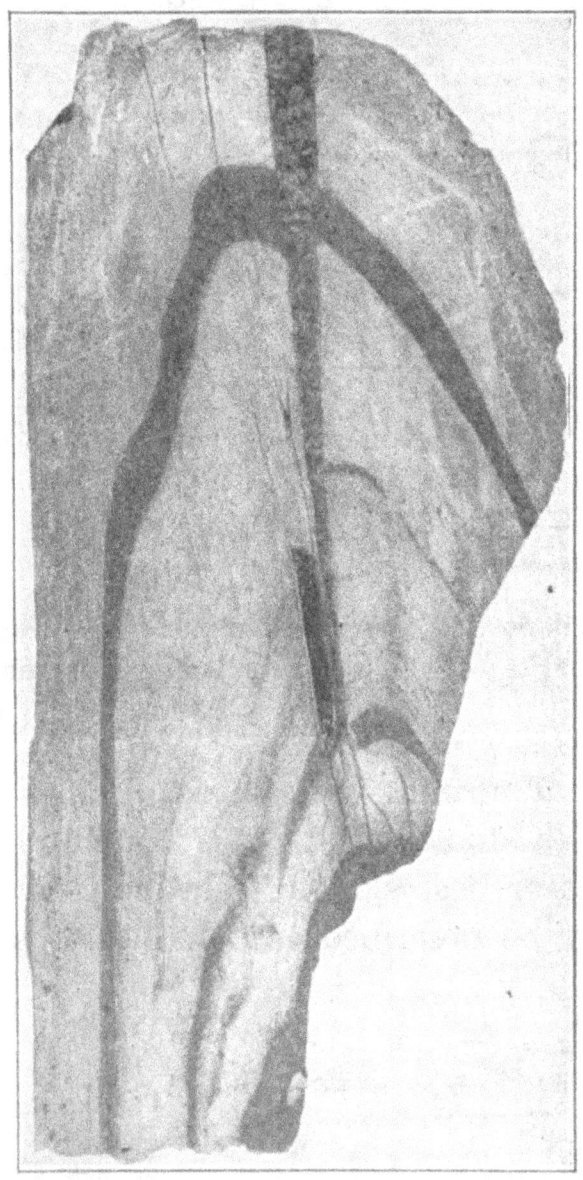

FIG. 10.—Specimen of folded slate, Black Hills, South Dakota, showing a fissure developed along the axis of the fold. *After Irving.*

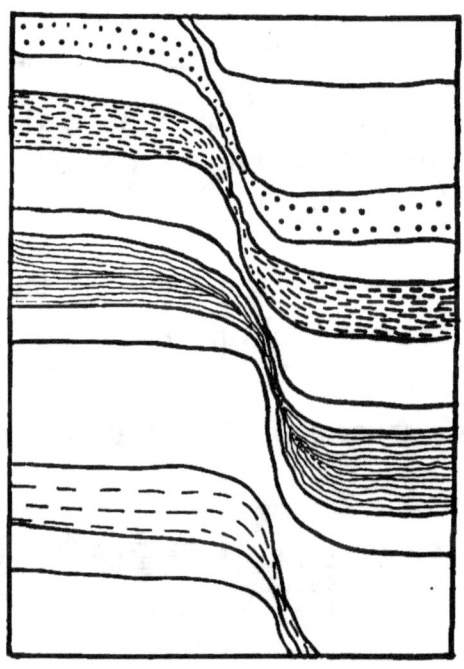

FIG. 11.—A flexure. *After Beck.*

both folding and faulting these expressions of stress are likely to be parallel.

Fractures.—As generally used, the term fracture denotes a break in a rock mass in importance intermediate between a joint and a fault, as these latter terms are generally used.[1]

Flexures.—A flexure is a sharp bend in a series of strata or in a rock mass, without the development of a continuous fracture, the result being a displacement similar to that of a monocline. Flexures readily pass into faults.

Faults.—A fault is a fracture through a rock mass the opposite walls of which have moved past each other, the word indicating lack of correspondence between opposite walls.

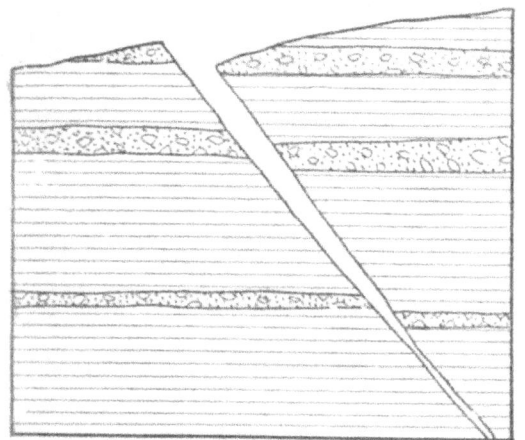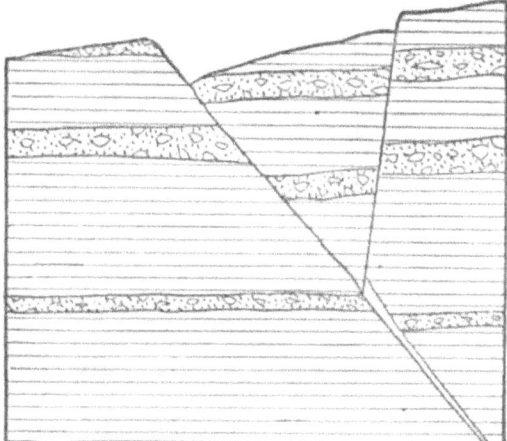

FIG. 12.—Normal faults: a recent fault near the surface, and the same with the hanging dropped down. *After Beck.*

The strike of a fault is the direction of a horizontal line within the plane of the fault; the dip of a fault is the angle between the plane of the fault and a horizontal plane.

The distance measured on the plane of a fault between the new positions of two points that were originally opposite is called the total displacement, which distance may, for purposes of calculation, be considered as the hypothenuse of a right triangle, whose sides represent the horizontal and the vertical movements of which it is the resultant.

The terms throw and heave as commonly used refer respectively to the vertical and horizontal distances between the new positions of originally opposite points in a fault the movement along which was directly down the dip; by offset is com-

[1] J. E. Spurr, "Geology Applied to Mining," p. 177.

3

monly meant the horizontal distance (perpendicular to the strike) between the two portions of a faulted vein, without regard to the direction or amount of total displacement.

In a great majority of faults the total displacement is not directly down the dip, but is the result of both horizontal and vertical movements.

A normal fault is a fault along which the hanging wall has moved downward on the foot-wall, and is the natural result of a drawing apart of the two rock masses, the least supported mass slipping downward on the mass having the larger base.[1]   Normal faults are commonly the effect of gravity, and due to tension,

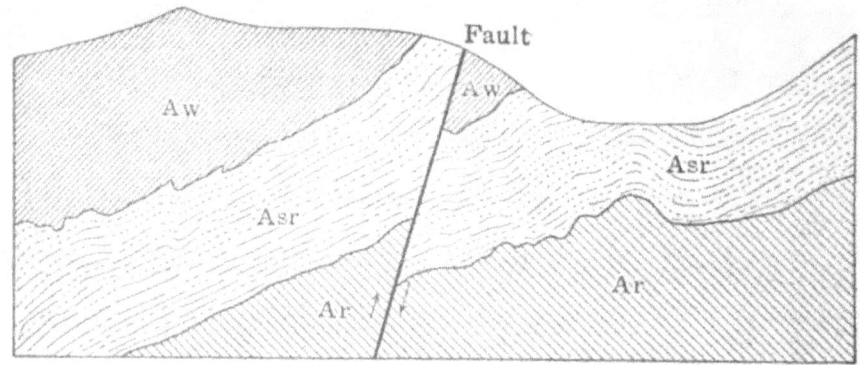

Fig. 13.—Reverse fault, Coeur d'Alenes, Idaho.   *After Ransome.*

although they sometimes result from compressive stresses. Faults due to tension commonly find expression in a simple fracture and are not accompanied by sheeting.

A reverse fault is one along which the hanging wall has moved upward on the foot-wall, which is commonly considered as having been forced under the overlying mass by compression.   Such faults are less likely to be simple fractures than normal faults, and are frequently accompanied by parallel sheeting, or by a folding of the strata or rock masses that form their walls.

Faults due to torsional stresses are commonly accompanied by differential movement, the displacement over one part of the fault being greater than over other parts, resulting in a tilting of the faulted block.   Such faults frequently follow curved lines of strike, and are often branching.

[1] J. F. Kemp, "Ore Deposits," p. 21.

Faults are sometimes referred to as strike faults, or as dip faults, according to their direction as compared to the strike or dip of enclosing strata or associated veins.

A series of approximately parallel faults of similar class and displacement are called step faults; if of opposite displacement, they are called compensating faults.

Two faults, or two systems of faults, parallel in strike but inter-

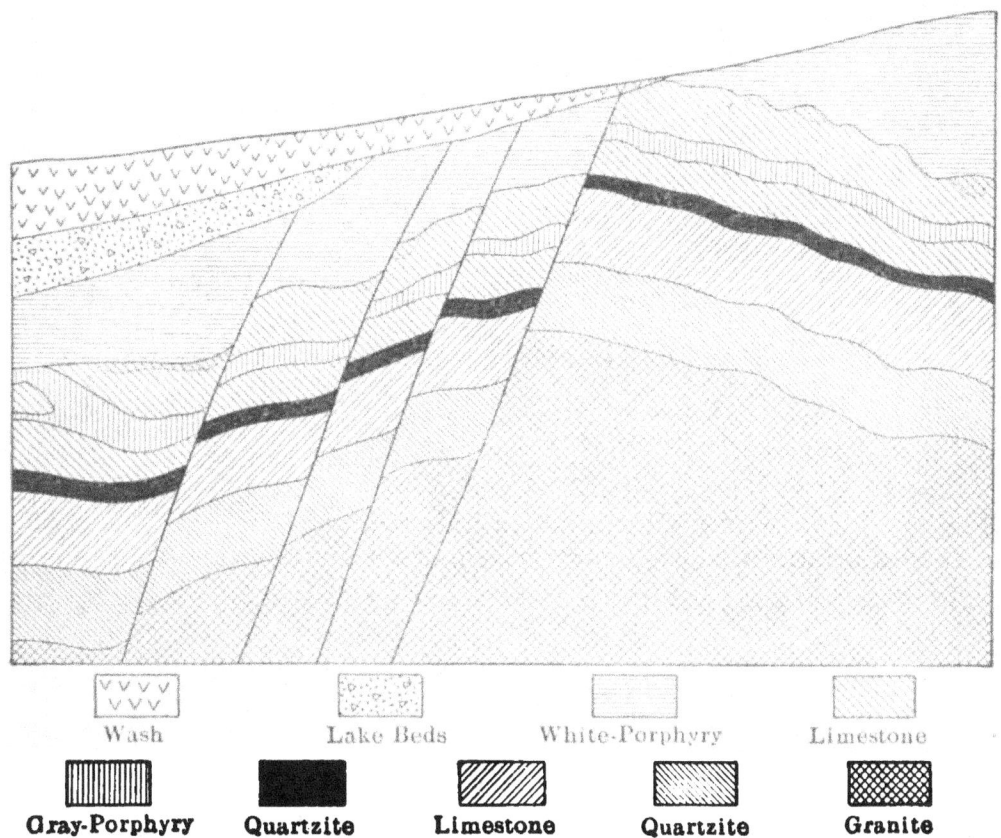

Fig. 14.—Section perpendicular to the strike of a series of step faults at Leadville, Colorado: The total displacement of a fault that extends for a long distance through the region is here distributed among several parallel faults. *After Emmons and Irving.*

secting in dip, are occasionally formed by the same compressive or torsional stresses, and are known as conjugate faults or fault systems.

Faults formed at slight depth below the surface are likely to be less regular and less persistent than deep-seated faults. The occurrence of many small, irregular fractures that become less in number and more regular in direction as depth is gained, is indicative of faulting under light load and at slight depth; a further indication of shallow formation is the presence of friction breccia along a fault rather than the pasty gouge that is characteristic of deep-seated movements under heavy loads of overlying rocks. The upward branching of a fault, or a marked change in dip, is a sign of shallow dislocation. Surface detrital matter is occasionally found in a fault filling, conclusively proving a shallow depth at the time of faulting.

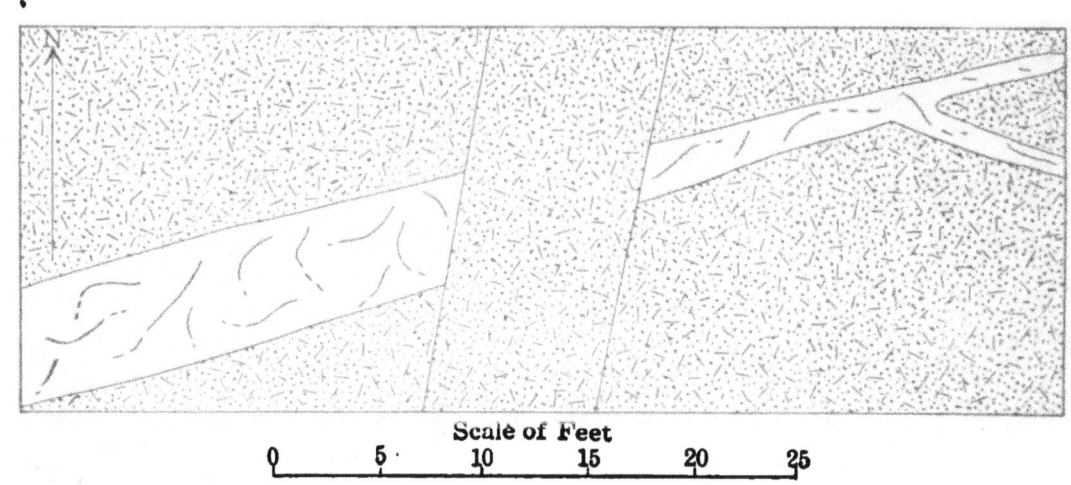

Scale of Feet

FIG. 15.—Horizontal sketch plan of a part of the Mizpah vein, Tonopah, Nevada, showing probable compensating faulting. *After Spurr.*

Exploration for the Continuation of a Faulted Ore-body.—Where an ore-body has been cut off by a fault the discovery of its continuation beyond the fault is a problem of the greatest importance. Numerous rules have been formulated to aid in a search for the continuation of a faulted ore-body, but it is safe to say that the great variety and complexity of the results of faulting render such rules valueless in most cases.

Where faulting dislocates a non-tabular ore-body that is contained in a homogeneous rock, the problem is insoluble, unless the fault carries a drag or trail of ore mixed with the fault filling.

The direction of the striations on the walls of a fault, in simple cases, indicates the direction of movement, the deeper ends of the furroughs being pointed in the opposite direction to the movement of the opposite walls. Occasionally an ore-body is deformed where faulted, the deformation pointing in the direction of displacement and passing into a drag or trail through the fault filling.

Where a tabular ore-body or vein is faulted, the problem is likely to be simpler, as the recovery of any portion of the vein beyond the fault will lead to the discovery of the continuation of the ore-shoot. Where the enclosing rock is composed of a series of strata a knowledge of the succession of the strata is of the greatest assistance in the calculation.

The cases are rare where complete data for the calculation of the total displacement are available, as, for instance, where a fault cuts a dike that itself cuts a series of known strata, or an older fault, or any combination of intersections that permits the recognition of a point rather than of a plane on both sides of the fault.

In an endeavor to locate the continuation of a faulted ore-body the striations or slickensides should be studied, and also the fault filling for possible fragments of ore, or of country rock other than that of the immediate walls of the fault; a survey, map, and sections should be made showing the dip and strike of both the fault and of the ore-body, if of tabular form, and of every known stratum of the containing rock, igneous contact or dike, older fault or vein, or any distinguishing mark that may be recognized on both sides of the fault. The application of descriptive geometry, or trigonometry, to this data will result in all the information regarding the displacement that the situation renders possible; complicated formulæ, like rules of thumb, are of no assistance in the solution of such problems.

It is seldom that the direction and amount of total displacement may be measured directly, more frequently it is susceptible of calculation from related functions, and frequently it is not determinable until further exploration discloses the necessary data.

A comparison with known faults in the same district is frequently of great help in the consideration of an unknown displacement, and an apparently complicated fault system once worked out may become a relatively simple problem when again met with underground.

Post-mineral Fissures.—A line of weakness once established is likely to continue as such through long periods of time, the original gouge seam or zone of attrition material being preserved and offering a plane of easy relief to subsequent stresses; where a fissure has been completely healed by mineralization, the resulting vein of brittle quartz or other minerals is likely to be less resistant to fracture than the tougher enclosing rock. It is common, therefore, to find evidence of post-mineral movement along veins.

Post-mineral movement is likely to crush the ore and to mix it with waste to such an extent that its value is materially reduced and also to render the country rock loose and likely to cave during mining operations; not infrequently, post-mineral movement so complicates the structure as to render exploration difficult, or unremunerative.

A favorable effect of post-mineral fissuring, either along or across a vein or deposit, is in permitting the access of surface waters, which are thereby given opportunity to form secondary enrichments, as will be taken up in a later chapter.

A usual effect of post-mineral movement along a vein is the formation of false walls, or fissures carrying gouge, that cut the vein obliquely; these mask the part of the vein behind them, and also hide the junctions of branch veins. False walls are frequently the cause of losing a vein in development, the tendency being to follow their well-defined fissures and to leave the vein to one side. Where such movements are known to have taken place, frequent cross-cutting may be necessary to make sure that the whole of the vein has been exposed.

Pre-mineral Fissures.[1]—Gouge-filled fissures appear to be unfavorable to ore deposition, probably chiefly because the pasty gouge does not permit the ready passage of mineralizing solutions, and also because the continual movement along such

[1] This subject is fully taken up by Mr. F. L. Ransome in *P. P.* 62, U. S. G. S., p. 120.

fissures tends to close any minute channels that are permitted to form. Such fissures once formed probably persist as lines of weakness and movement for long periods, and the occurrence of a gouge-filled fissure in connection with an ore-deposit is no proof that it is of later formation than the ore.

It is frequently seen in mines that such fissures act as efficient dams, the passage of solutions through them being as difficult as the circulation of solutions along them; they may be considered, therefore, as having frequently determined the limits of ore deposition through the impounding of mineralizing solutions.

The fact that an ore-body or vein is cut off by a gouge-filled fissure is no proof that a fault has displaced the ore-body or vein, the continuation of which may never have existed beyond the fissure. It is often difficult to prove whether such a fissure is of later formation, and has faulted a vein, or whether it is older than the vein, which ends upon reaching it. The proof of a fault is, of course, the continuation of the vein beyond it; this, however, is the object of the search. That such a fissure is a post-mineral fault may be indicated by the faulting of associated beds or dikes, or by a drag or trail of ore through the fault filling.

If such a gouge-filled fissure is older than and limits the ore, this may be indicated in a filled fissure, by a closing of lines of crustification against the fault, equivalent bands being connected, or by a change in the vein filling upon approaching the fault, or by a branching or widening of the vein upon approaching the fault.

A gouge-filled fissure older than the ore may through recent movement exhibit the characteristics of a fault that has displaced the vein, though in reality limiting it.

The Expression of Faults in Topography.—The more recent a fault, and the greater the difference in resistance to erosion between the rocks of its walls, the more likely is it to find expression in the topography; a fault of great displacement that finds no expression in the topography is probably not of recent origin. Faults may in some cases be recognized at the surface by features other than the lack of correspondence between their walls, such as a fault scarp, a saddle in a ridge, or the course of a canyon, or through the outcropping of a fault breccia that has become silicified and so rendered resistant to erosion. In most cases, however, erosion is the dominant feature in controlling the topography, and structural features are rarely represented at the surface, except through relative resistance to erosion.

The Importance of Areal Geology.—The surface affords a complete section of the geological features of any district, and in all but the simplest occurrences a geological map of the surface and a few vertical sections are invaluable guides in the examination or exploration of any district. All significant outcrops of rocks, dikes, beds, veins, faults, shear zones and so forth should be determined and their strikes, dips, and elevations recorded on a large scale map, from which the data may be referred for study to a horizontal plane. Contour maps are useful in such work, but in most cases the expense of contouring is not justified, and the elevation of significant readings will be found to be sufficient.

A horizontal section thus prepared, with, perhaps, several vertical sections, will indicate through lack of correspondence of strata or other criteria, the existence and location of faults that otherwise might not be suspected. If the surface observations are abundant, they often afford data from which the displacements of faults may be calculated.

A geological map of the surface results in dividing a district into fault blocks, conditions within each of which may be considered as constant, but which must be expected to change upon passing from one fault block to another. The importance of dividing a district into fault blocks will be appreciated on considering that a prospect may be situated within a few hundred feet of an important mine, but actually separated from it, perhaps, by a fault thousands of feet in throw, or in a formation that is apparently similar, although actually unrelated, to the ore-bearing rock.

Exploration and Development.—In the opening of any property two distinct objects should be kept in mind—the development of known ore-bodies, and the exploration for further deposits.

Where payable ore has been found, it should be followed with a view to the development of ore and also to gain knowledge in regard to its occurrence, each ore-shoot being opened individually without the testing of theories. In general, the time to go slowest in following an ore-body is when it shows signs of giving out;

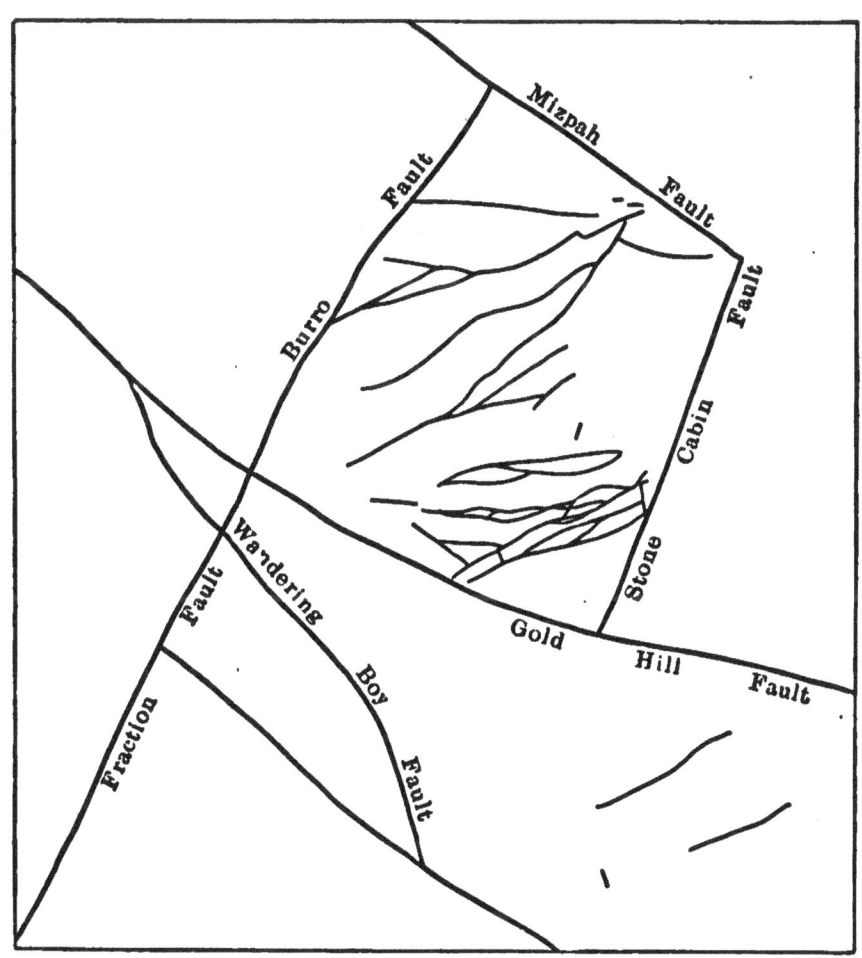

Fig. 16.—Map showing the fault blocks at Tonopah, Nevada; the payable veins occur in the earlier andesite and are most prominent in the fault block bounded by the Burro, Mizpah, Stone Cabin, and Gold Hill faults. *After Spurr.*

every detail should then be studied and recorded before the rock has been dirtied by further blasting, as frequently small stringers or gouge-filled fissures will be found to lead from one ore-body to another.

In the exploration for further ore-shoots, however, the major features of the distribution of ore should be borne in mind, and the general trend of the deposit should be cross-cut thoroughly. Exploration should be confined at first to cross-cutting the known ore-bearing horizon or zone of mineralization—perhaps a sedimentary bed favorable to the deposition of ore, a zone of crushing or brecciation, or a certain horizon known to be the most favorable for secondary enrichments. Within these broad zones or horizons considered favorable to the existence of ore, the minor features, such as stringers of ore, low-grade ore, sheeted zones and so forth, may be considered important. Upon cross-cutting the apparent trend of the deposit at the horizon considered most favorable to the existence of ore, the most promising stringer or seam should be followed along its strike for an appropriate distance, where another cross-cut is in order.

The plan to be followed demands detailed study in each individual case, and a state of mind is necessary that is receptive of new impressions as the work progresses. Preliminary exploration is frequently entrusted to a practical miner, when as a matter of fact, such work constitutes perhaps the most important field of the trained geologist.

CHAPTER III

STRUCTURAL FEATURES OF ORE-DEPOSITS

Veins, Lodes and Ledges.—Many definitions have been advanced and many limitations advocated in the use of these terms. The following definitions appear to follow the best usage.

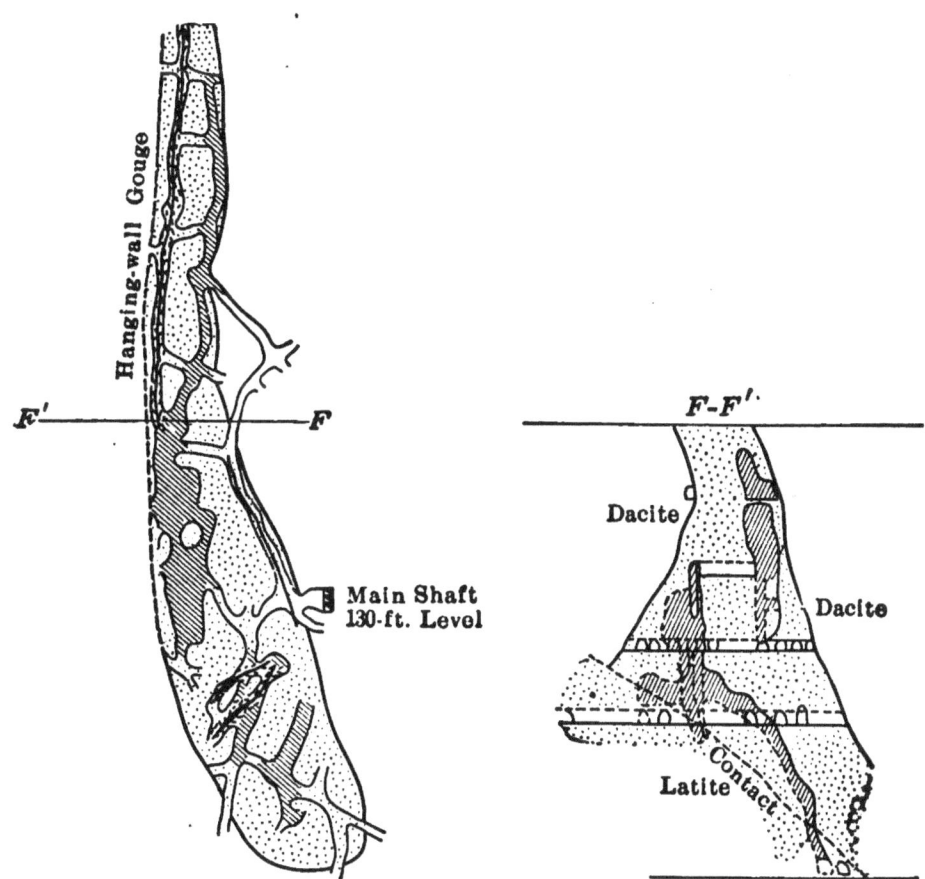

Fig. 17.—Plan and section of the Combination ledge, Goldfield, Nevada, showing the irregular occurrence of the ore within the ledge. *After Ransome.*

Fissure Veins.—A fissure vein is a mineral mass tabular in form as a whole, though often irregular in detail, filling or accompanying a fracture or series of closely set and intimately related parallel fractures in the enclosing rock, the mineral mass having been formed later than both the country rock and the fracture, either through the filling of open spaces along the fracture or through chemical alteration of the adjoining rock.[1]

Lodes.—A lode is a zone of fissuring that contains roughly parallel mineral masses of the general type of fissure veins, usually connected by cross veins and mineralized breccias to such a degree that over certain portions the whole width constitutes a single ore-body.

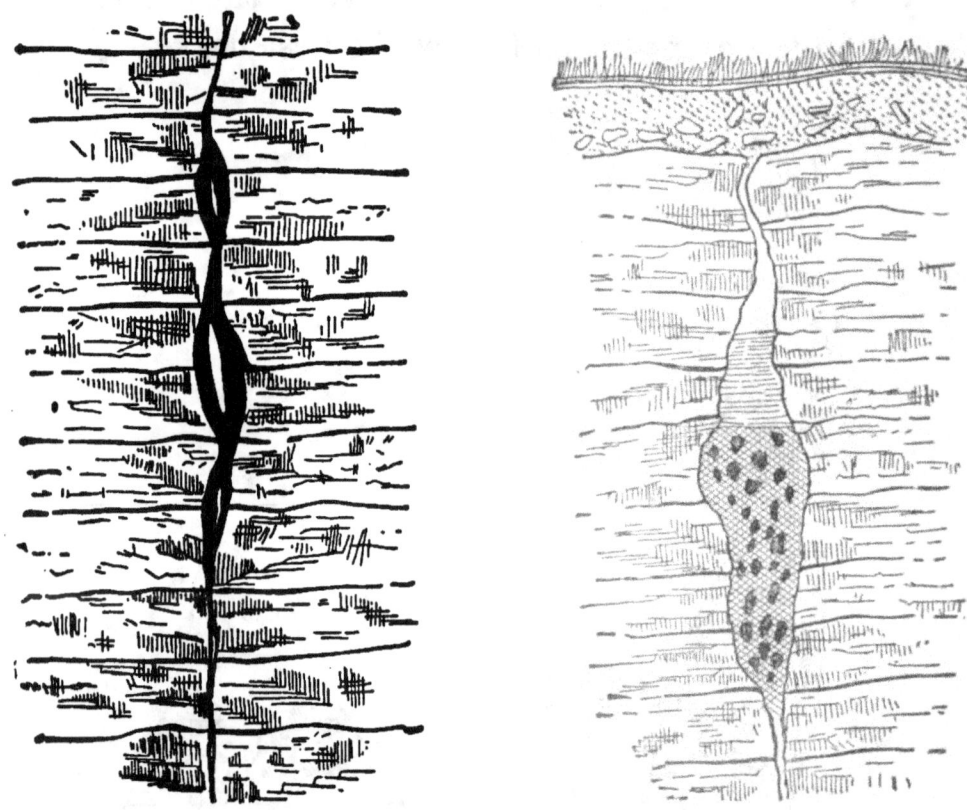

FIG. 18.—Gash veins, Upper Mississippi Valley; the black shading indicates galena. *After Kemp.*

A ledge is an irregular mass of altered rock, containing ore bodies, the alteration of which is due to and characteristic of the action of mineralizing solutions.[2]

Gash Veins.—A gash vein is a vein of superficial character, widest near the surface and narrowing to extinction in depth. Gash veins are usually the results of solution and deposition

[1] Waldemar Lindgren.
[2] F. L. Ransome.

along joints or small fissures by surface waters, and are of secondary origin. The term has been less correctly applied to lenticular deposits that, prominent at the surface, die out in a similar manner in depth. The usual occurrence of gash veins is in sedimentary rocks. The typically short length of this type of vein should not lead to an expectation of continuity in depth, even if the origin is not suspected from study of the outcrop.

Bed Veins.—A bed vein is a vein that follows a bedding plane of an enclosing sedimentary rock, less frequently a plane between

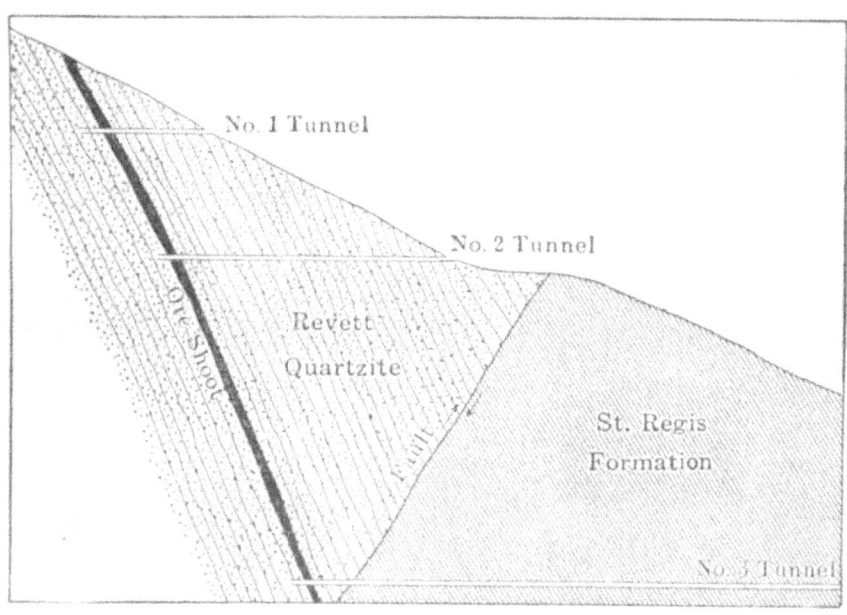

Fig. 19.—Section of a bed vein (copper) in the Snowstorm mine, Coeur d'Alenes, Idaho. *After Ransome.*

layers of volcanic rocks. Bed veins are commonly thought to be less persistent than veins that cut across the strata of enclosing rocks; many cases are known, however, where bed veins are both persistent and contain important ore-shoots. Blanket vein is often used as a synonym for bed vein, but actually refers to a horizontal or nearly horizontal position only.

A bed vein in unaltered rocks is sometimes distinguished with difficulty from a stratum whose mineralization was contemporaneous with its own deposition. In the examination of an unaltered bed, an opinion may usually be based upon the closeness with which the mineralization follows the minute bedding planes as well as the larger bed divisions; if the mineralization follows the larger bedding features only, and does not penetrate the relatively solid intervening strata, it is probably later in origin than the bed itself. Fragments of country rock or the presence of cross or branching stringers are definite signs of a later origin. Furthermore, a mineralization that is not relatively continuous through a certain stratum, but is transferred to strata above and below without corresponding lateral extent, is later than the containing bed.

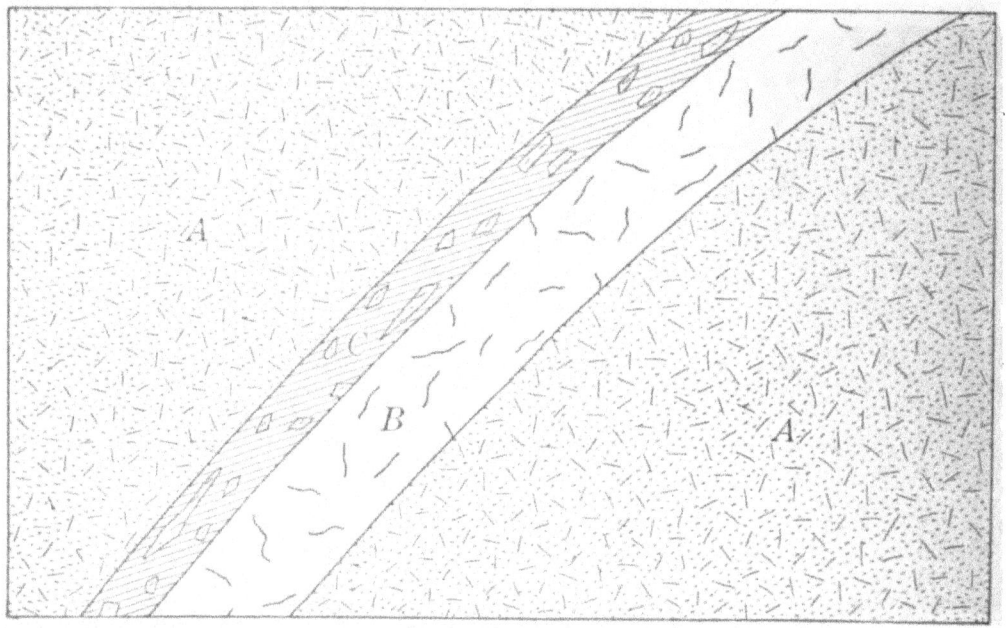

Fig. 20.—Diagrammatic section of the Tonopah Extension vein, Tonopah, Nevada, showing a compound vein. *A*, Altered wall rock; *B*, typical white vein of the earlier andesite period, containing silver sulphides and carrying values of several hundred dollars per ton; *C*, black, jaspery quartz of later introduction than the original vein of which it carries fragments—the value of the black quartz and fragments varies from twenty to thirty dollars per ton. *After Spurr.*

In the Coeur D'Alenes, Idaho,[1] a majority of the auriferous quartz veins are of the bed-vein type. These veins in some qlaces occur singly, but more commonly they occur in groups, individual veins being separated by a few inches or a few feet of slaty country rock. An overlapping arrangement is common, one vein gradually pinching out while a parallel vein between

[1] F. L. Ransome, *P. P.* 62, U. S. G. S., p. 141.

adjacent beds becomes correspondingly thicker. Although certain individual veins persist for hundreds of feet without cutting across the planes of stratification, such crossings may be observed here and there, and cross-cutting stringers of quartz that link neighboring veins are numerous.

Compound Veins.—A compound vein is a vein that has been reopened after mineralization, and again mineralized, containing,

FIG. 21.—Open pit of the Treadwell mine, Douglas Island, Alaska, showing an albite-diorite dike that has been brecciated and impregnated with ore.   *After Spencer.*

perhaps, two sets of ores. A frequent case is that of a vein composed chiefly of barren or low-grade filling that has received its valuable mineral upon being reopened.

Contact Veins.—A contact between two rocks occasionally offers a line of weakness to fracturing that later mineralization may transform into a contact vein. Such a vein is rarely regular over long distances, unless its position along the contact be accidental, due to faulting. Veins of secondary minerals are likely to form along a contact where one wall offers an impervious barrier to migration of solutions and so induces precipitation along its course through impounding.

Veins Along Dikes.—Dikes of intrusive rocks, after solidification, are likely to remain lines of weakness along which new fractures readily form; veins frequently follow dikes, either along one wall or through the mass of the dike itself.

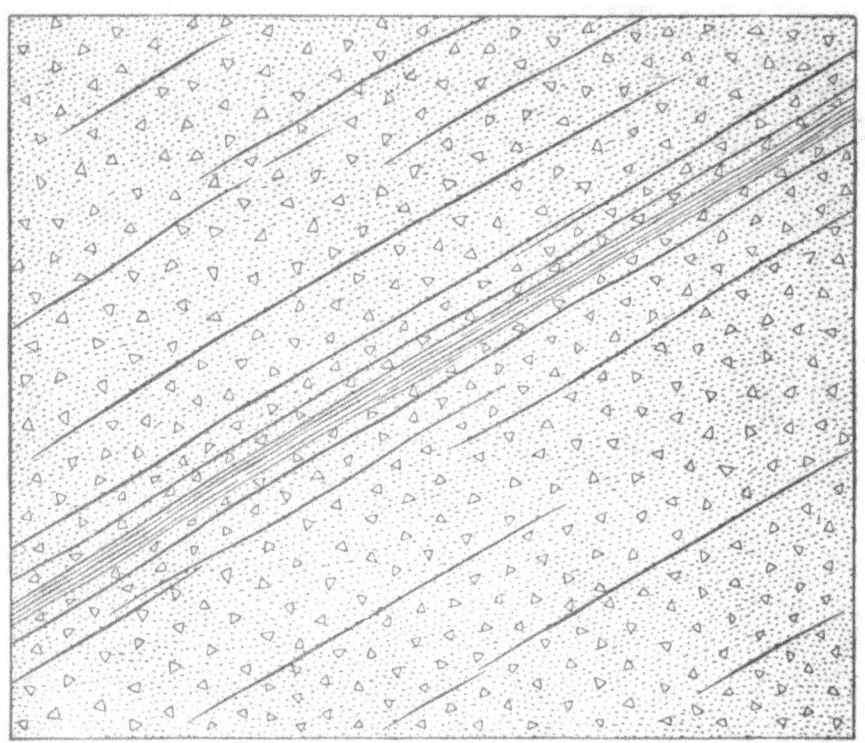

Fig. 22.—Section of the Howard lode, Cripple Creek. Colorado. showing a sheeted zone; the medial portion of the lode is closely sheeted, the sheeting planes becoming gradually farther apart in the foot- and hanging-walls. *After Lindgren and Ransome.*

At Cripple Creek, Colorado, many of the veins follow phonolite or basic dikes.

Lodes Along Sheeted Zones.—A sheeted zone is a series of closely set parallel sheeting planes; they frequently afford channels for mineralizing solutions and so become mineral lodes. Such lodes are commonly not as persistent as fissure veins, but mineral deposits may follow them for long distances, the mineralization being found sometimes in one and sometimes in another set of sheeting planes, following, perhaps, an overlapping arrangement. Sheeting planes are commonly parallel to the main fissures, and lodes are frequently made up of a mineralized fissure with one or more mineralized sheeting planes nearby. There is usually no evidence of movement along sheeting planes.

Where there has been much replacement along a lode the sheeting planes are likely to be obscured, and the term is usually applied, therefore, to lodes that carry their minerals in a scanty gangue. A typical lode along a sheeted zone is made up of narrow parallel veinlets separated by slabs of country rock.

Frequent cross-cutting is advisable in the exploration of a mineralized sheeted zone.

AT CRIPPLE CREEK, COLORADO,[1] sheeted zones consist of a number of narrow, approximately parallel, fissures, which collectively form lodes ranging from a few inches up to 50 or 60 ft., or rarely, 100 ft., in width. Within such wide belts of fracture, however, two or more zones of concentrated fissuring may usually be distinguished that lie close enough together to be mined as a whole. In other words, the very wide sheeted zones are compound sheeted zones. As a rule, the fissures are mere cracks, showing no brecciation, slickensiding, or other evidence of appreciable movement of the walls; there are some notable exceptions to this statement, but the movement of one wall past the other has probably in few instances exceeded 1 or 2 ft. In general, the sheeted zones are from 2 to 10 ft. in width. Among the numerous and important lodes that properly come under the designation of sheeted zone, several structural varieties may be distinguished, which are sometimes exhibited in different parts of the same lode. A common form is that characterized by the presence of two main parallel fissures, usually 3 or 4 ft. apart, accompanied by less regular and less persistent fractures in the intervening and adjacent rock. In another common type of sheeted zone the parallel fissures are more numerous, and are spaced with some regularity. There is usually a medial portion of the lode that ranges from a

[1] Waldemar Lindgren, *P. P.* 54, U. S. G. S., p. 160.

few inches to a foot or two in width, within which the rock is
divided into a large number of very thin plates by fissures often
less than an inch apart; this band of intense sheeting is accompanied on both sides by parallel fissures that are spaced farther
and farther apart, so that the sheeted zone as a whole merges
gradually into the country rock.   There are a large number of
sheeted zones in breccia and in granite that are composed of
many parallel or nearly parallel fissures, but that differ from the
type just described in the absence of a well-defined medial zone
and in the rather less regular character of the fractures.

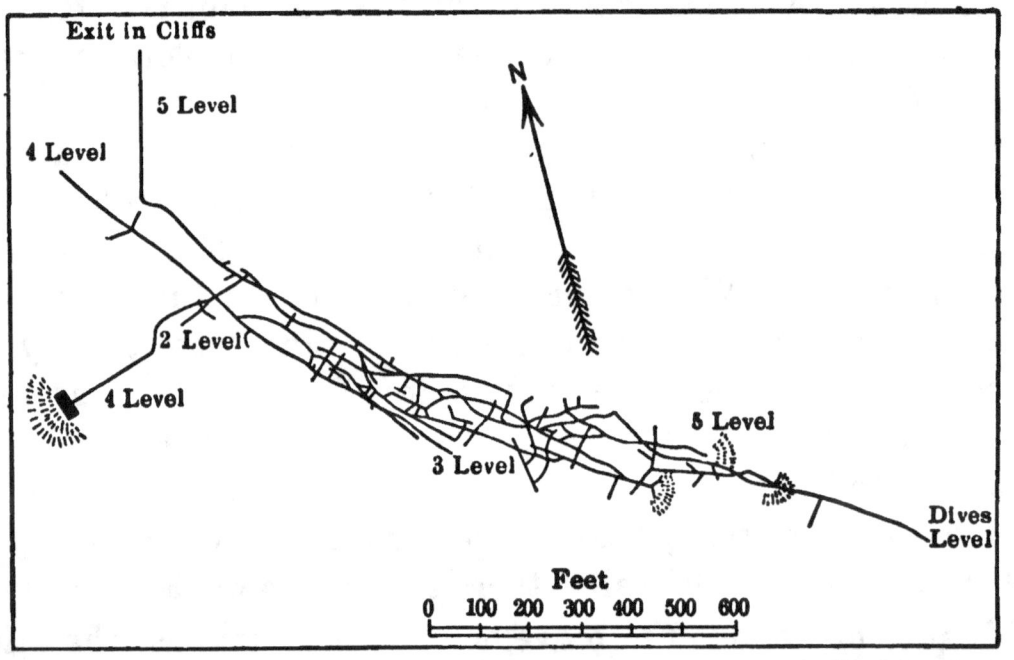

Fig. 23.—Sketch plan of the underground workings of the North Star
mine, Silverton, Colorado, showing the irregularity of mineralization of a
stringer lode.   *After Ransome.*

Stringer Lodes.—A lode in which the mineralization has followed a net-work of irregular curving fissures that have no
general parallelism among themselves, but follow a general trend
similar to that of a sheeted zone is known as a stringer lode.
This type of lode rarely has definite walls, and the ore may be
encountered at any point within the general zone of fissuring; as
with sheeted zones, frequent cross-cutting is necessary in the
exploration of this type of deposit.

Fault Lodes.—A zone of faulting in which mineralization has taken place irregularly through the crushed and comminuted fault material is known as a fault lode. This type of deposit, where the values are irregularly distributed either with or without a scanty gangue, is not uncommon in the areas of old schists in the desert region of the Southwest. These lodes are difficult to follow, and their exploration in advance of actual mining is usually unprofitable.

The Mineralization of Joints.—A strongly jointed rock offers

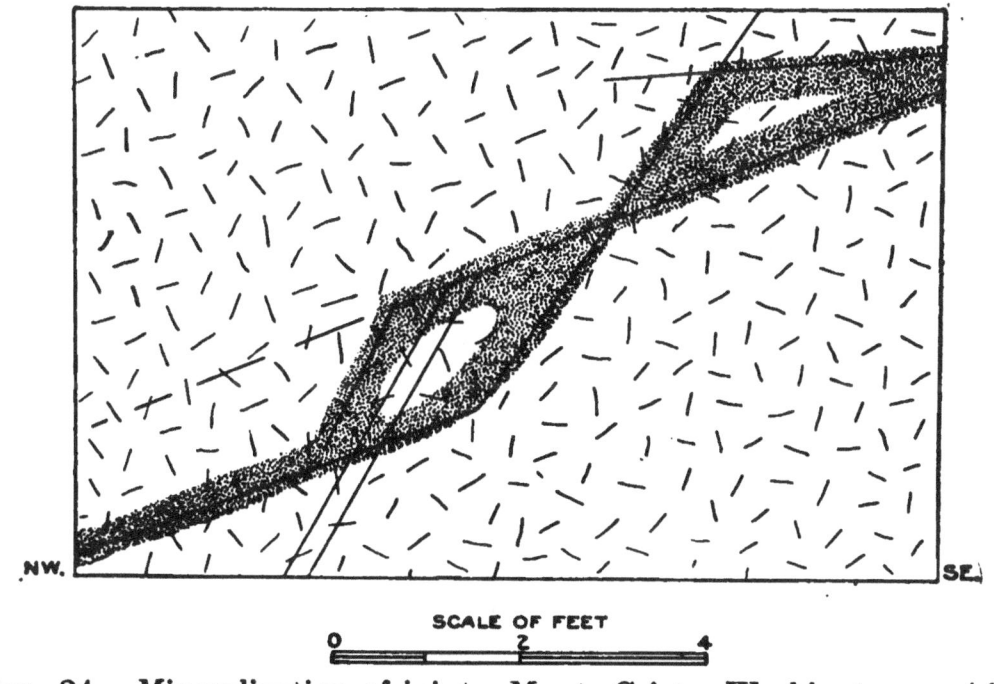

SCALE OF FEET

Fig. 24.—Mineralization of joints, Monte Cristo, Washington. *After Spurr.*

many lines of weakness to fissuring stresses, and not infrequently a fissure that is well-defined in depth is dissipated through the joint planes upon nearing the surface. A mineralization that has taken place at relatively shallow depth, therefore, not infrequently follows the joint planes and is distributed among them as a system of reticulated veinlets. Such disseminations in depth are likely to coalesce into well-defined lodes. A system of mineralized joint planes is commonly the result of surface agencies; less frequently it is the result of a primary mineralization at shallow depth in a region where erosion has been slight.

Breccia Lodes.—A zone of shattering in which the mineralization has cemented or replaced the brecciated mass of angular fragments and comminuted material is known as a breccia lode. This type of deposit is commonly irregular, the mineralization varying with the amount and degree of brecciation. In extreme cases the brecciation may have been out of proportion to the

SCALE OF FEET

FIG. 25.—Mineralization of joints, Monte Cristo, Washington. *After Spurr.*

quantity of mineralizing solutions, and have dissipated these solutions through so large a mass of rock that the resulting ore body is very low in grade.

Shear Zones.—A zone of incipient fissuring or shearing that has been mineralized by impregnating solutions, commonly by replacement, and to a less extent, perhaps, by the filling of open spaces, is known as a shear zone. The passage of solutions through such zones is probably slow, and ample time is afforded for mineralization by replacement. While genetically a shear zone may be of any size, the term as generally used is applied to large, low-grade deposits.

Stockworks.—An area through which numerous veins traverse the rock in all directions, forming a net-work through mutual intersection, is known as a stockwork. In typical cases the individual veins are small and are considered collectively as a deposit.

Stocks.—A stock is a deposit of irregular form due chiefly to replacement of the containing rock.[1]

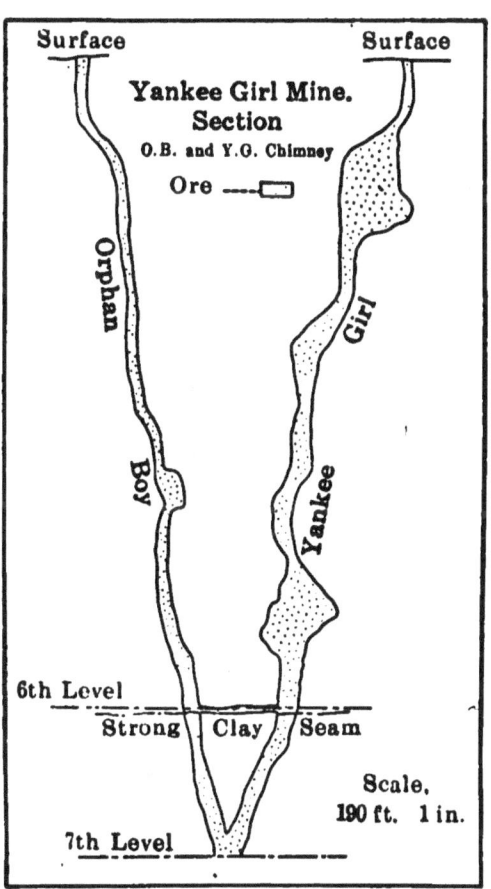

FIG. 26.—Ore chimneys in the Yankee Girl mine, Red Mountain, Colorado. *After Schwartz.*

Pipes or Chimneys.—Certain deposits take the form of a pipe or chimney, and have marked vertical continuity, with very subordinate horizontal dimensions. It is difficult to conceive of a fissure whose only important dimension is approximately vertical, and these deposits were probably formed at the intersection of fissures that throughout the remainder of their lengths

[1] Beck-Weed, "Nature of Ore Deposits," p. 51.

did not permit the passage of mineralizing solutions, and so failed
to be emphasized in connection with the main deposit; other
deposits of this form appear to be due to the mineralization of
fumarolic vents. Certain pipes in limestone appear to have
been formed by solution, probably along a fracture, or intersection of fractures: these are likely to be quite irregular in form.

AT THE BASSICK MINE, CUSTER COUNTY, COLORADO, an ore-deposit that varies from 20 to 100 ft. in horizontal diameter has been

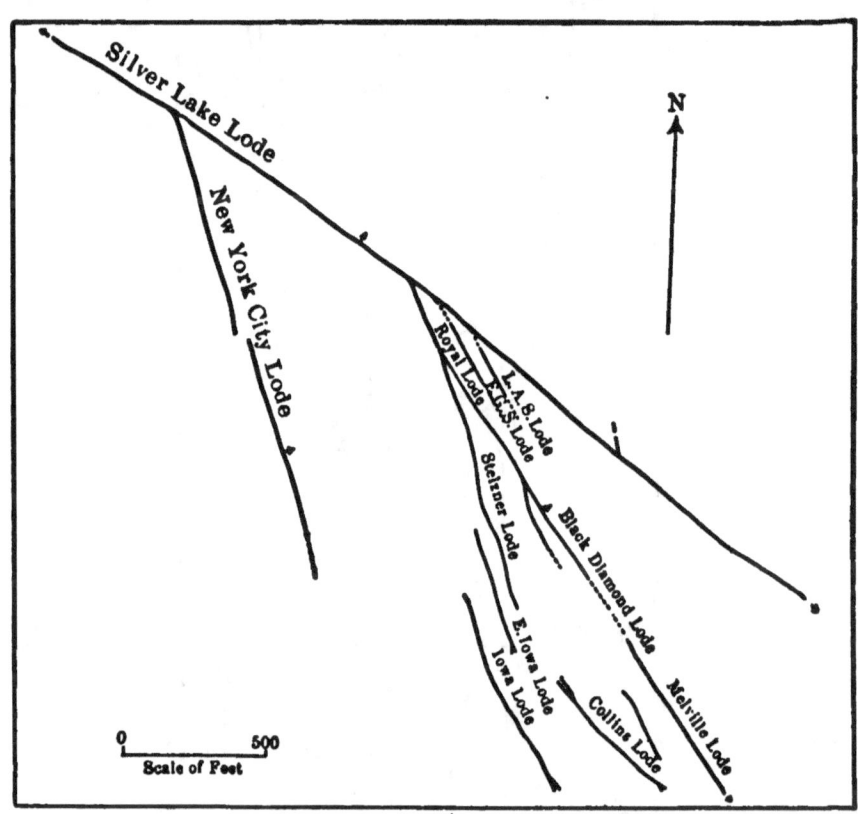

FIG. 27.—Plan of the productive lodes west of Silver Lake, Colorado,
showing a branching vein system; the lodes farthest south are an example
of an overlapping system. *After Ransome.*

followed to a depth of over 800 ft. This chimney occurs in a
volcanic neck and the ores have formed in concentric layers
about boulders of the volcanic agglomerate.

Branching Veins.—Many veins send off branches into their
hanging- or foot-walls, usually into the former; beyond such
branches veins frequently lose in either width or value. A vein
that branches along its strike may or may not unite farther on,

but, in general, the branching of a vein is likely to indicate the dying out of its fissure. If possible, therefore, development work should follow the direction in which the branches are converging. While no definite rule may be formulated, branch veins are likely

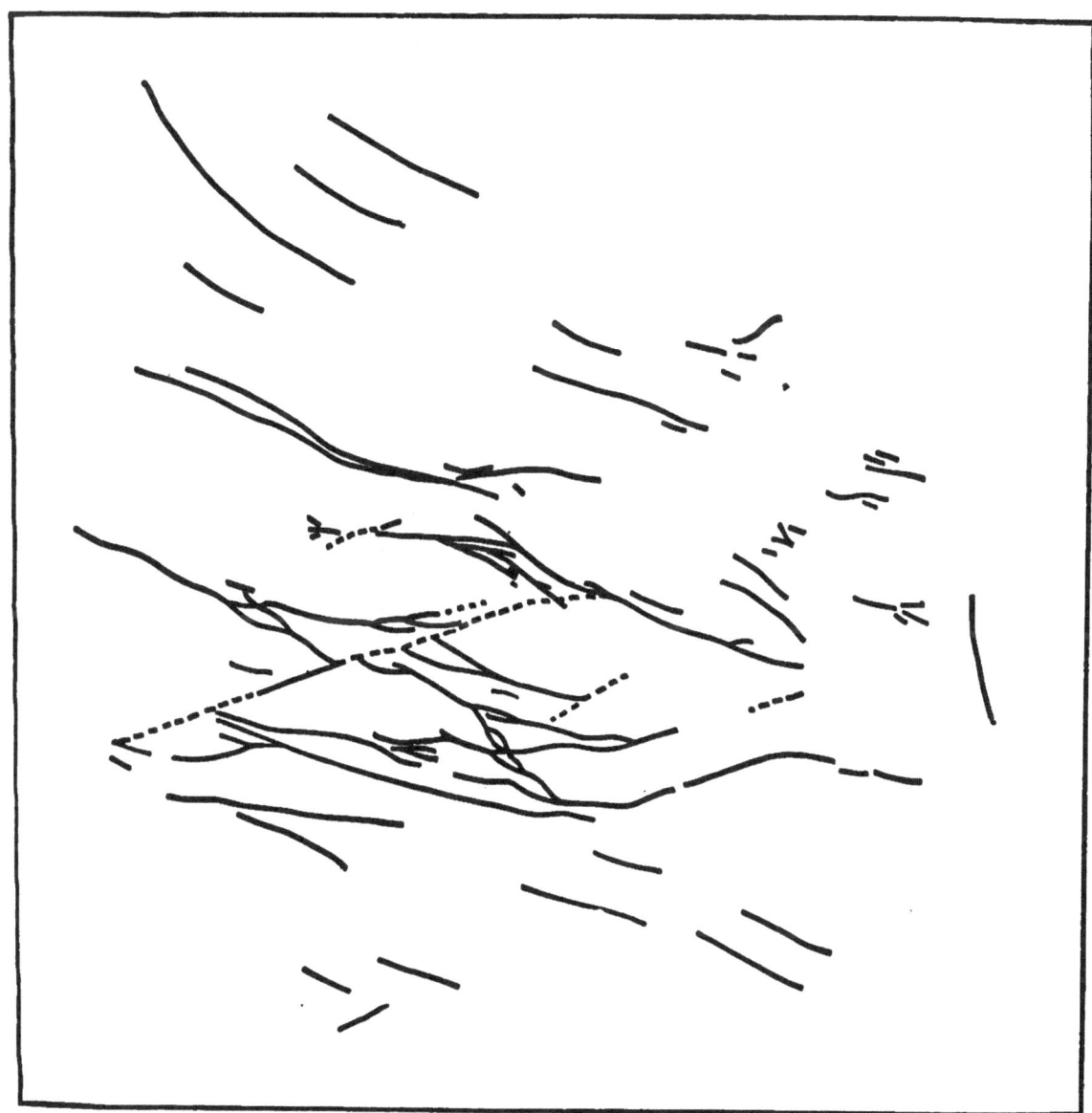

Fig. 28.—General sketch of the lode system of the Upper Harz, Germany, showing a linked-vein system. *After Beck.*

to diminish in both size and value with distance from the main vein, which usually formed the main channel of mineralizing solutions, and of which the branch veins were only lateral and dependant channels.

Linked Veins.—A system of veins that branch and reunite

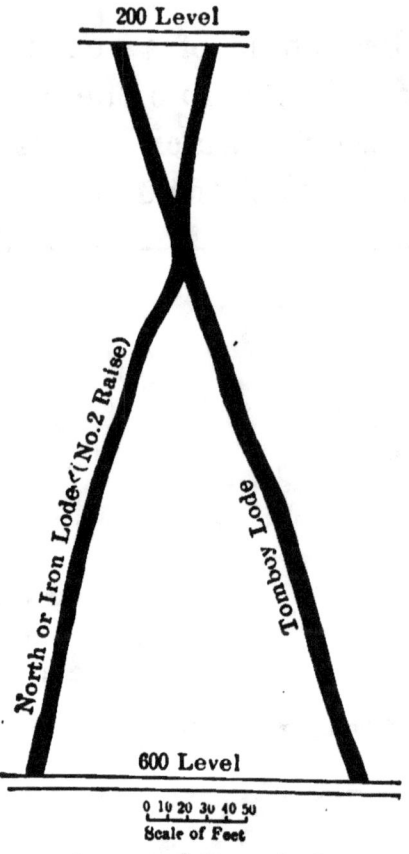

FIG. 29.—Section of the Tomboy and Iron lodes, Silverton, Colorado. *After Herron and Ransome.*

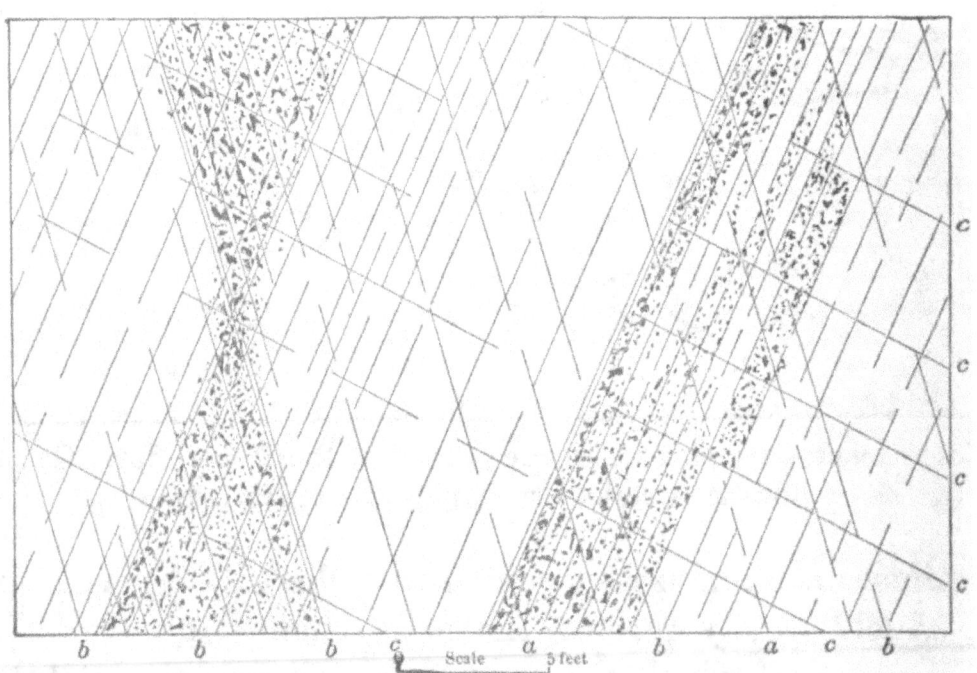

FIG. 30.—Intersecting, or conjugate, joint systems, mineralized, Encampment, Wyoming. *After Spencer.*

without crossing one another is called a linked-vein system. Where such a relation exists between veins, the arrangement appears likely to be persistent.

Conjugate Veins.—Fractures produced by compressive stress through homogeneous rocks are likely to form systems parallel in strike but opposite in dip, known as conjugate fractures; where mineralized these become conjugate veins, of which one dip is usually emphasized while the other is subordinate, or exists as a simple fissure only. Conjugate joint systems are of common occurrence. · The intersection of two systems of intense jointing appear to be favorable *loci* for ore deposition.

Overlapping Veins.—It often happens that the stresses producing fractures do not find expression in a single persistent fissure, but in a number of more or less closely spaced parallel. fissures that partly overlap one another, forming a step-like system. Such a system is known as a system of overlapping fissures. Overlapping veins are probably the result of fissuring stresses that were exerted along a line other than the line of least resistance of the rock mass; while the lines of easiest cleavage of the rock were followed over relatively short distances, the final expression of the stress was along the average strike of the individual fissures taken as a whole, and parallel to the simple fissure that would have formed had the rock been homogeneous.

Many vein systems follow overlapping fractures, and this arrangement probably gave rise to the miner's rule to cross-cut "into the hanging" or "into the foot" upon losing a vein, the overlapping being either persistently to the right, or persistently to the left, in most cases. An overlapping vein system is also spoken of as having an arrangement *en échelon*. Blind veins, or veins that do not outcrop, frequently belong to an overlapping vein system, one of whose members outcropped and so led to the discovery of those that did not.

Systems of Related Veins.—A careful plotting of the veins of any mining district will often indicate a certain relationship among the payable fissures as regards strike, dip, or distribution. A clustering of veins about a particular intrusive mass is frequently observed, and the important veins in many districts maintain a fairly parallel alignment. Mr. J. M. Boutwell,[1] has determined that at Bingham, Utah, over 84 per cent. of the payable fissures strike between N.5° E. and N. 43° E., and that

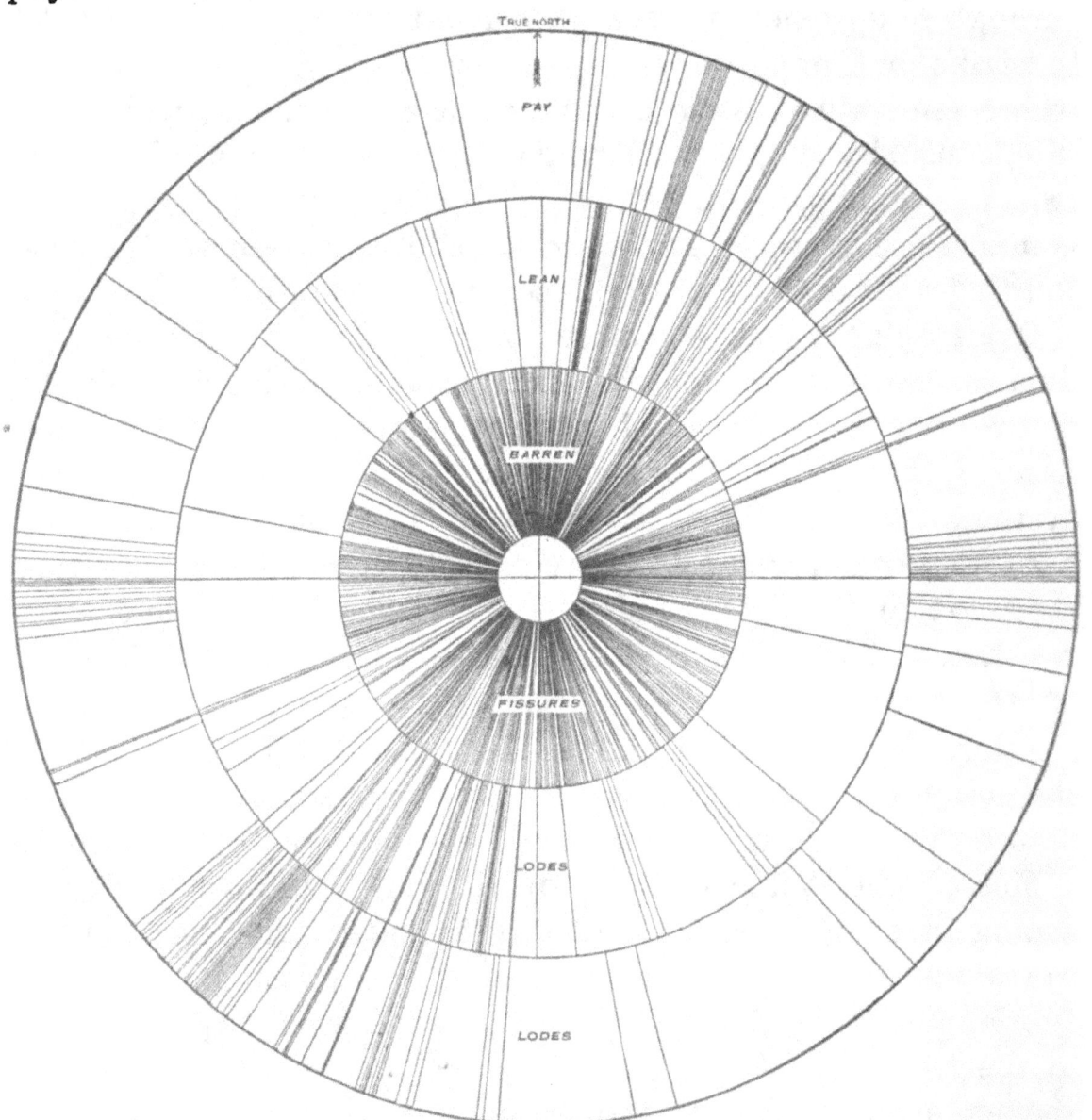

FIG. 31.—Diagram showing the trends of the barren fissures, the lean, and the payable veins and lodes of the Bingham District, Utah. *After Boutwell.*

over 80 per cent. of them dip between 40° and 90° NW. In this district all fissures, barren, lean and payable, taken together, strike about equally in all directions, but the northeast fissures

[1] *P. P.* 38, U. S. G. S., p. 363.

appear to have been those that afforded channels for solutions at the time of mineralization. A striking instance of the parallelism of pay fissures is exhibited in the Bald Mountain District, South Dakota.

A roughly radial distribution of veins about a common center has been noted in certain districts, and while well established in

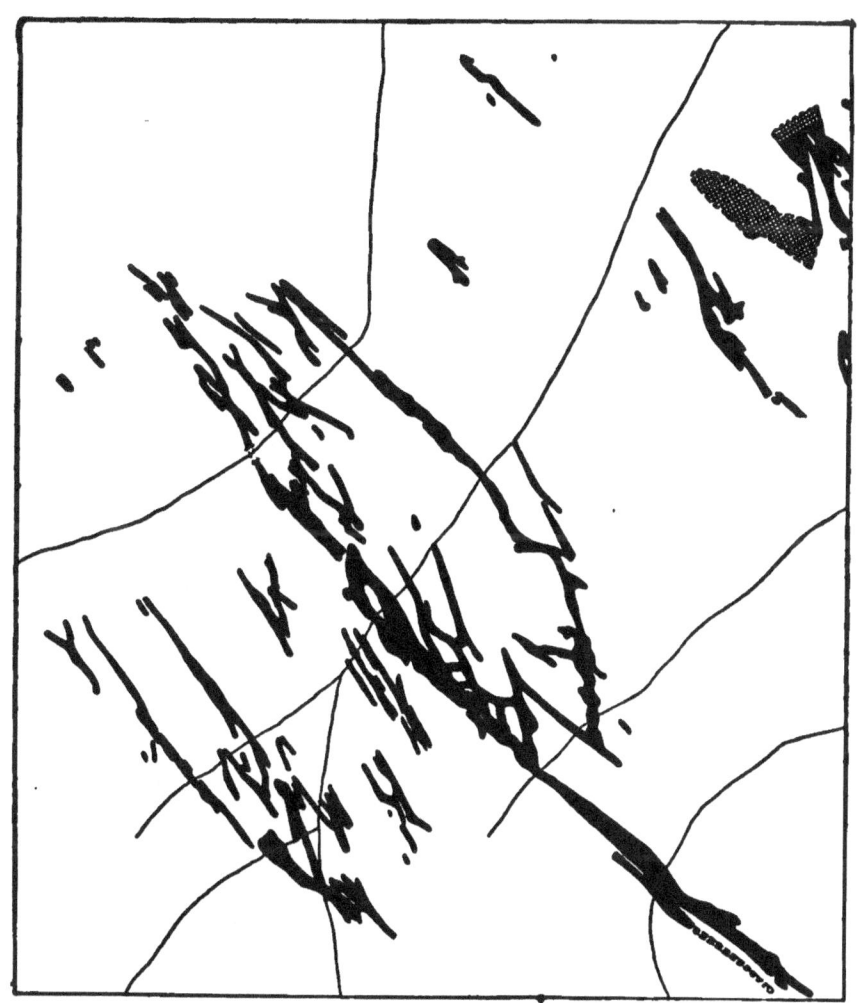

Fig. 32.—Map of the principal ore-shoots of the Bald Mountain area, Black Hills, South Dakota, showing the marked parallelism of the payable fissures. *After Irving.*

these districts, it is of rare occurrence. Cripple Creek, Colorado, affords perhaps the best known example. The veins at Cerro de Potosi, S. A., are said to follow a roughly radial distribution.

While the recognition of a definite vein system is a valuable guide in exploration, the vein systems that have been cited indi-

FIG. 33.—Plan of the principal veins of the Cripple Creek District, Colorado, showing a roughly radial distribution about a common center. *After Lindgren and Ransome.*

cate rather more clear relationships than commonly obtain among the veins of most districts.

In districts where the veins vary greatly in strike and dip it is easy to construct relationships that do not exist, and thus arrive at misleading conclusions.

The Persistence of Veins in Depth.—There is a distinct relationship between the length and depth of veins; in general, long, strong, wide veins persist in depth and short, non-continuous, irregular and weak veins die out at no great distance beneath the surface, perhaps to be followed by similar and roughly parallel veins in depth. While ore-bearing fissures are commonly not important as faults, there is a relation between the amount of throw and the length and depth of a fissure; the greater the throw, the greater are likely to be these dimensions.

In the consideration of individual veins that have pinched in depth, a decision will rest upon the behavior of the vein in the parts already explored. If the proved length of the vein is considerably greater than the depth attained, and if the displacement along the vein is known to have been more than slight, the chances are good that the fissure will persist in depth. This has long been recognized by miners, who consider well-developed slickensides an indication of persistency in depth. If a vein has been subject to local pinches above, it may be safely assumed that it will open out again with deeper exploration. In general, it may be said that the behavior of a vein in horizontal exposures is likely to be roughly duplicated down its dip. A possible change in the country rock should be borne in mind, however, as fissures and veins are likely to change markedly in structure upon passing from one rock into another.

A vein that is about to die out is likely to split into several diverging and diminishing stringers, and gradually to become lost in the country rock. This termination is more often noted in horizontal directions than in the dying out of a vein in depth, probably for the reason that few veins are followed in depth to their complete extinction.

The probabilities in regard to the ultimate depth of fissures and veins, while of theoretical interest, have little or no practical significance. The depth at which the plasticity of rocks under great pressure no longer permits the existence of openings is far below the depth of possible mining operations.[1]

The Relation Between Depth and the Number and Character of Veins.—In most mining districts veins are more numerous at and near the surface than in depth. Veins are much more likely to come together in depth than to divide as they go down, and many subordinate and weak veins die out altogether within relatively slight depths below the surface. It is usual, moreover, for veins to become more regular in strike, dip and width in depth than they are nearer the surface, although they are commonly narrower in their deeper than in their upper parts. It appears probable that fissuring stresses are limited in their effect at considerable depth to a few fissures or to a single fissure, whereas near the surface the lesser burden of superposed rock permits a dissipation of the stresses among a number of more irregular fractures.

Mineral Veins Follow Fissures of Small Displacement.—Fractures through rock masses may be said to vary from planes of straining and incipient fracture to faults of great throw. Mineralization may take place along zones of strained rocks, and while such conditions are structurally favorable to replacement, the passage of solutions must be slow at best. The other extreme, that of faults of great throw, appears to be unfavorable to the passage of solutions and mineralization.

Fractures through rocks are rarely plane surfaces; they commonly follow curves over long distances, and through short distances are subject to many irregularities in strike and dip. The movement of one wall of a fissure past the other wall produces much finely ground material known as gouge, which in faults of large displacement commonly fills the fissure tightly, cementing the rock fragments and preventing any free flow of solutions.

[1] Prof. Van Hise gives as his opinion that at depths of from 10,000 to 12,000 meters the weight of the overlying rocks is too great to permit the hardest rocks to retain their form.

Faults of small displacement are, in general, channels favorable to the ready passage of solutions. The irregularities of the walls not having been planed off, slight movement of one wall past the other brings the concavities opposite one another, thus forming open spaces; furthermore, the strained rocks bordering such fissures not having been ground up into gouge, permit the solutions to circulate through them and present conditions favorable to replacement. Most payable veins follow fissures of small displacement, although there are notable exceptions to this rule.

The Influence of Country Rock on Vein Structure.—A fissuring stress that produces an efficient circulation channel in one rock

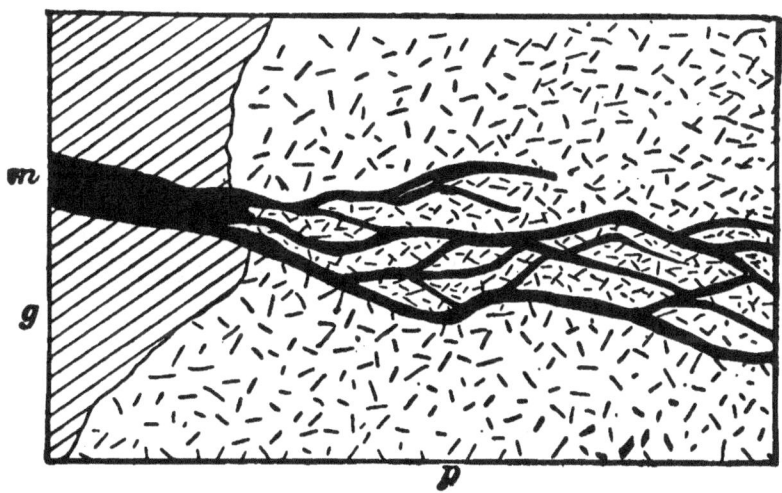

Fig. 34.—Scattering of the Gottlob Vein, Freiberg, Germany, upon passing from gneiss into quartz-porphyry. *g*, Gneiss; *p*, quartz-porphyry; *m*, vein. *After Beck.*

may be too great, or too small, to produce a like result in a rock of different character. The size, regularity and character of fissures vary according to the physical properties of the enclosing rocks, such as homogeneity, friability or plasticity.

A vein that traverses more than one country rock, therefore, is likely to vary greatly in structure in the several rocks. Fissures of small displacement, which are the sort commonly followed by mineral veins, may be strong and clean cut through a massive rock, but may fade out, and become lost in a plastic rock, where flowage or distortion is likely to take up the movement as a slight fold, without the production of a fracture.

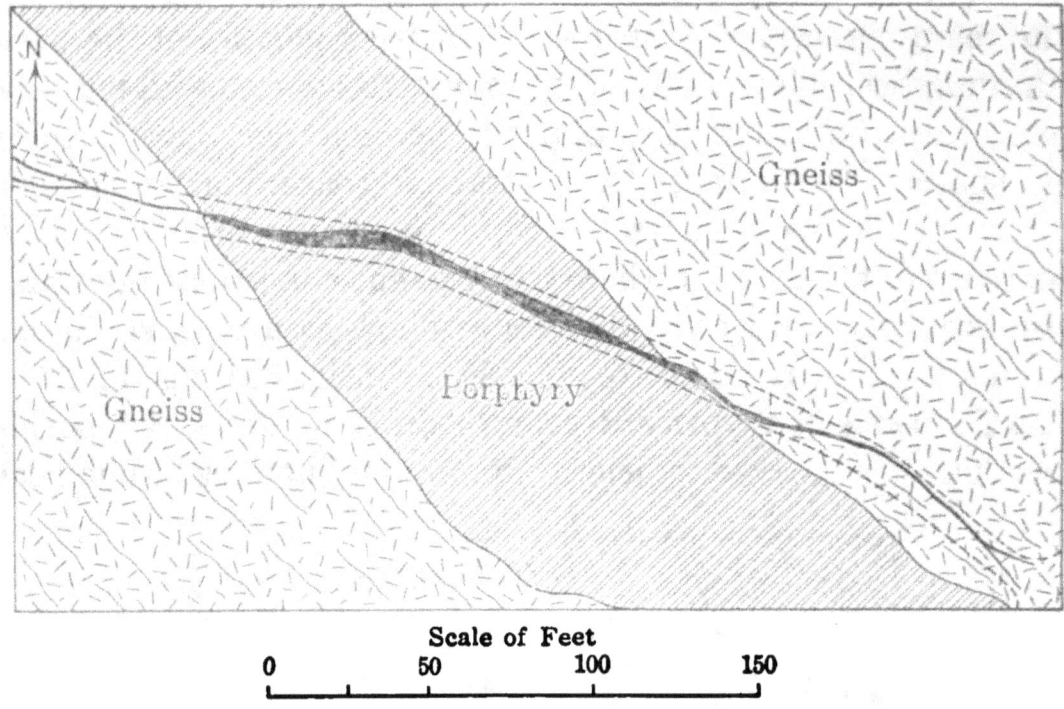

Scale of Feet

0   50   100   150

FIG. 35.—Horizontal plan of the vein in the Maine mine, Georgetown, Colorado, showing thickening of the vein in passing from gneiss into the harder porphyry. *After Spurr and Garrey.*

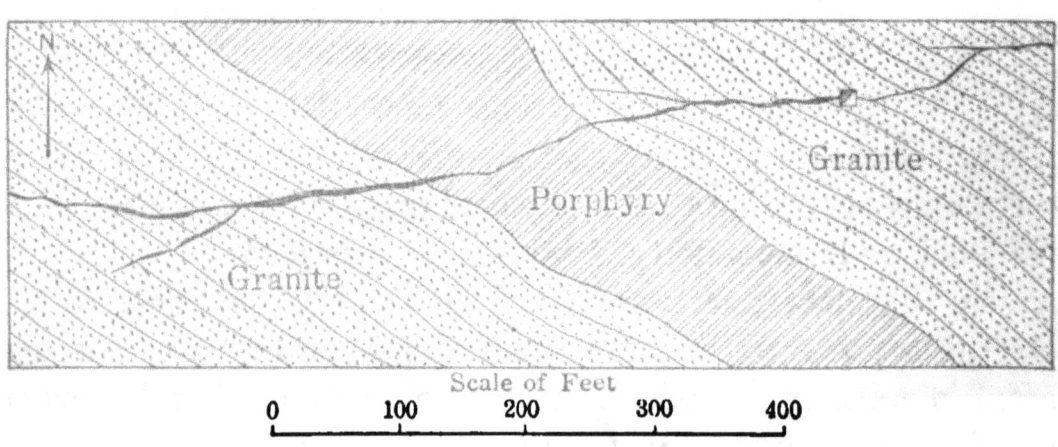

Scale of Feet

0   100   200   300   400

FIG. 36.—Horizontal plan of a part of the Seven Thirty vein, Georgetown, Colorado, showing diminution in the size of the vein in its passage from granite into porphyry. *After Spurr and Garrey.*

A vein that is persistent and regular in a homogeneous rock upon passing into a brittle, unyielding rock, or into a rock mass that is seamed with closely spaced joints, is likely to become dissipated, and its mineralization scattered through so large a

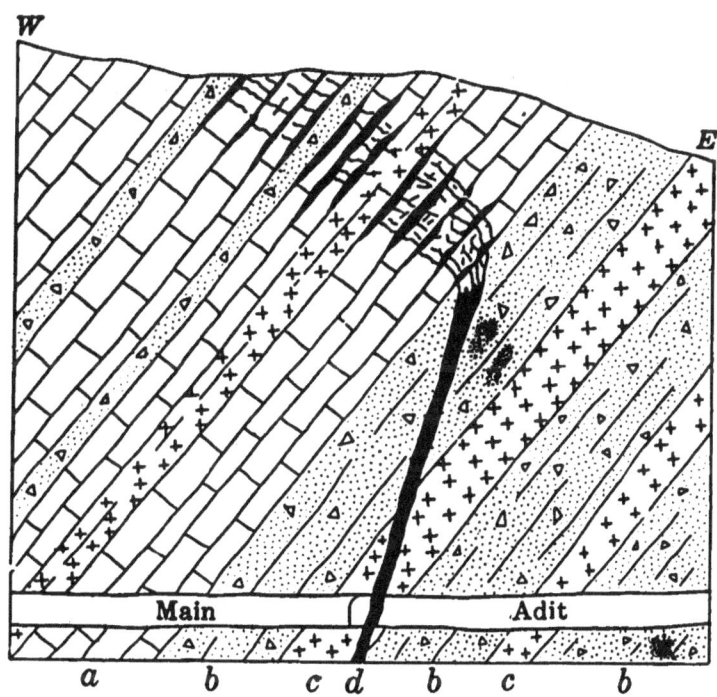

FIG. 37.—Diagrammatic section of the Sheba ore-body, near Unionville Nevada, showing the behavior of the vein in different rocks. *a*, Limestone; *b*, tuff; *c*, porphyry; *d*, ore. *After Ransome.*

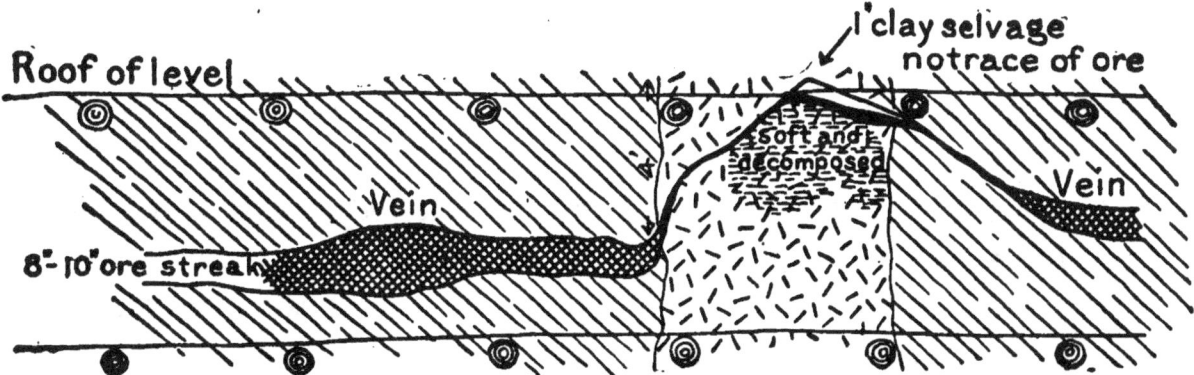

FIG. 38.—Behavior of vein at Neihart, Montana, upon passing from schist into amphibolite. *After Weed.*

mass that the resulting deposit is of too low grade to permit extraction.

Veins that are contained in a large mass of igneous rock, or other rock of relatively constant character, are likely to be more regular and more persistent than veins that traverse a series of bedded sediments.

Where it is seen, therefore, that a vein is to pass from the rock through which it has been followed into a rock of different character, decision should be reserved as to its probable continuity, unless abundant local evidence indicates that the change elsewhere in the district is without effect, or is favorable in character.

The behavior of a fissure at a contact between two rocks depends largely upon the angle at which it meets the contact.

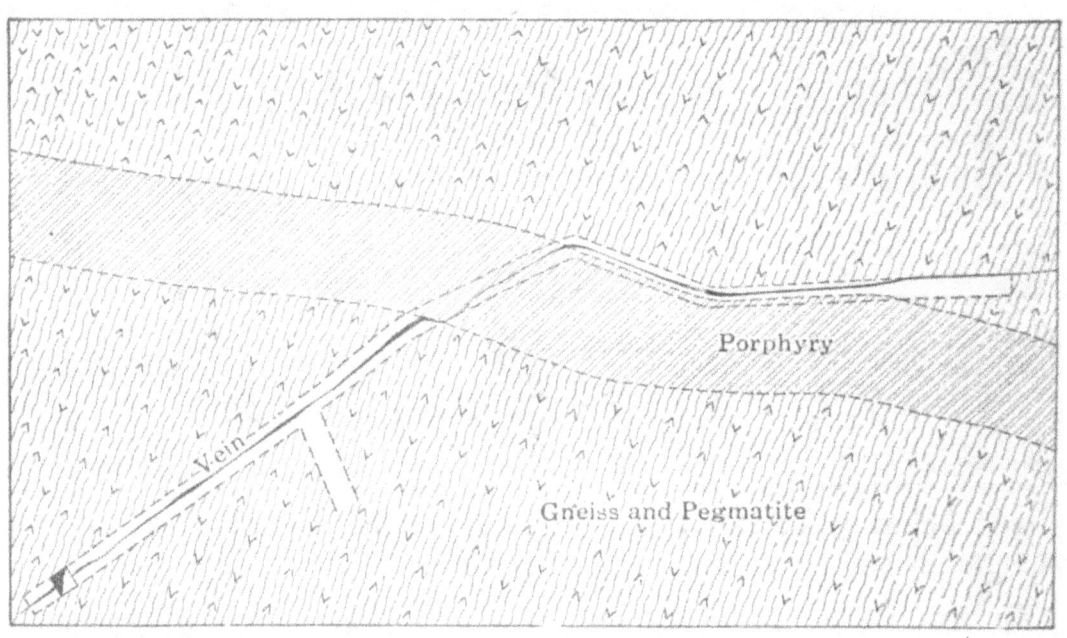

Fig. 39.—Sketch of a vein in the Gomer mine, Idaho Springs, Colorado, showing the deflection of the vein upon meeting a dike. *After Spurr.*

If the angle is acute, the fissure may be deflected along the contact for some distance, and, upon crossing it, continue with a strike parallel to its original course. If the fissure meets the contact at nearly right angles, it is likely to cross it without deflection, or, more rarely, to cease altogether.

In general, shales, schists and slates are more likely to absorb stresses without the production of definite or continuous fissures, and so to permit a dissemination of solutions and resultant mineralization, than are rocks of non-fissile structure. Glassy rhyolites, quartzites and other rigid and brittle rocks are likely to be shattered by stress, and a mineralization that is regular in another rock may upon encountering them become disseminated through a large brecciated mass.

Soft, plastic intrusives are likely to adjust themselves to stress without fracture, and lodes upon passing into such rocks are unlikely to continue far into them; in this connection it should be borne in mind that the softening of igneous rocks through the metasomatic action of mineralizing solutions takes place during mineralization, and is not a factor in the formation of the fissures; such alteration indicates rather than militates against the continuity of associated fissures.

AT NEIHART, MONTANA,[1] a series of steeply dipping metamorphic rocks, consisting of feldspathic gneiss, softer bands of more schistose rocks, and occasional tough amphibolites, is cut by an irregular intrusion of diorite, and all in turn are intruded by a rhyolite porphyry. Well-defined fissure veins cross all these rocks, meeting the metamorphic strata at nearly right angles. The veins vary somewhat in width and in the relative abundance of included rock fragments upon passing from one belt of feldspathic gneiss to another, and more markedly where they pass into the more schistose rocks, but the change is complete and abrupt where they meet the amphibolites; here the veins commonly narrow from a width of 7 or 8 ft. of good ore to a foot or more of barren gangue. Upon passing from the schist into the diorite, the veins invariably narrow, but continue well defined. In the rhyolite-porphyry the same fissures lose their compact character, and split into net-works of fractures through which the ore is dissipated; a well-defined vein upon passing into quartzite in the Big Seven Mine in this district splits into many small stringers.

The Distinction between Intercalated and Fissure Veins in Schistose Rocks.—This distinction is of great importance in the investigation of quartz veins through schists or gneisses, which frequently carry intercalated quartz lenses, or beds that are deceptive in appearance and not connected with the important mineralization of the district.

[1] W. H. Weed, *Trans.* A. I. M. E., XXXI, p. 637.

In a district where quartz lodes later than the schistosity **are** known to exist, and a quartz body conformable to the schistosity is found, an examination of the lode matter in thin section under the microscope will usually indicate whether the deposit is later than the schistosity, and perhaps connected with the important mineralization, or whether it is older than the mineralization and forms an integral part of the metamorphic series. Under the microscope minutely disseminated minerals, invisible to the un-aided eye, may be found which are characteristic of the commer-cially important veins, or, on the other hand, such minerals or inclusions may be lacking and the quartz under the microscope may show a schistose structure, indicating an origin contem-poraneous with that of the enclosing beds.

A careful investigation in the field will often disclose branching stringers, or minute veinlets, which indicate a later origin than that of the schists. Crustification is a clear indication of later origin.

Fissures and Lodes Formed Subsequent to the Principal Miner-alization.—In most mining districts the important mineraliza-tion was confined to one period, and where fissures and lodes of various origins are present, it is of the greatest importance to distinguish the commercially important series or system from the others. A careful plotting of all the fissures, whether ore-bearing or barren, will often indicate the prevailing trend of the pay fissures, and the general distribution of the pay fissures about some particular intrusive, or in some definite area or areas, may afford a basis upon which to decide whether or not a fissure is worthy of exploration. Lodes of different dates rarely carry the same vein fillings, and a distinction upon this basis is often practicable. The kind and degree of the alteration of the wall rocks along the veins affords another criterion.

CHAPTER IV

PRIMARY ORES AND THEIR DISTRIBUTION

Ore-deposits of commercial grade are local concentrations of great rarity when considered in relation to the area of unmineralized land surfaces, and they must therefore be considered as the products of exceptional and complex conditions. The data collected from a great number of individual deposits are sufficient to permit certain broad generalizations in regard to the conditions favorable to ore deposition which even numerous exceptions do not invalidate.

The broadest general relation of ore-deposits is with intrusive rocks, and while there are notable exceptions, the great majority of ore-deposits have a visible or closely inferred connection with intrusive rock masses.

Conditions that permit the dissipation of the ore-bearing vehicles do not permit the concentration of metals in deposits; the prime factor to consider in this connection is that structural conditions must be such as tend to concentrate and not to dissipate the ore-bearing solutions.

The disadvantage of extrusive rocks is at once apparent; any metals the magma may have contained are dissipated at the surface and are lost; with the exception of tiny vugs and cracks in quickly cooled flows, no concentrations of metals are known in extrusive rocks, and the lack of confinement and quick cooling apparently do not permit extrusives to exert mineralizing effect upon the rocks with which they come in contact. The close mineralogical similarity between extrusives and intrusives, and the great difference in mineralizing power is a proof of the ease with which magmas part with their metallic content, and of the mobility of the ore-bearing vehicles in comparison with the parent magma.

As surface flows present the extreme conditions permitting dissipation, so laccoliths probably represent the extreme conditions favorable to the concentrated action of ore-bearing vehicles. The intrusion of a laccolith causes a domal uplift of overlying strata, with attendant fissuring of these strata; the changes of volume during intrusion, cooling and contraction give rise to forces that reopen or keep open such fissures over long periods of time, and afford vents for the mobile constituents of the magma, which are the ore-bearing vehicles; in an extreme case the entire emanation from a laccolith may be conceived to pass through a single fissure.

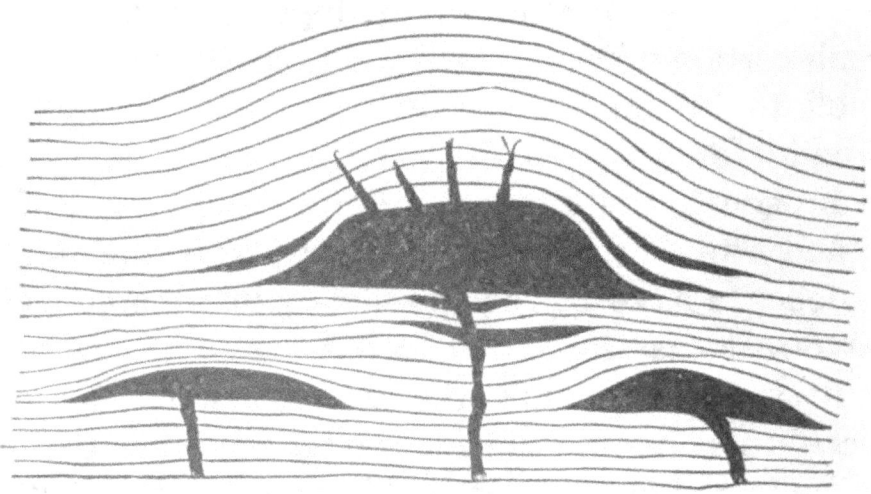

Fig. 40.—Ideal section of laccoliths. *After Gilbert.*

Between these two extremes of dissipation and confinement an infinite variety of intrusions are found in nature that give rise to an infinite variety of conditions under which ore deposition may take place.

In general, large bodies of igneous rocks do not carry ore deposits through their masses, the tendency being for the deposits to form in adjacent rocks, or in the first cooled parts or edges of the intrusive itself; the stresses that fissure adjacent rocks are absorbed in the still hot and plastic parts of the intrusive, but not infrequently shatter the already solidified peripheries and afford therein channels for the passage of the ore-bearing solutions and favorable *loci* for ore deposition.

Deep intrusives cut through and come in contact with many older rocks, and where these rocks exert precipitative action on the ore-bearing solutions occur other favorable *loci* for ore deposition.

In general, prerequisites for ore deposition are conditions that permit the escape of the ore-bearing vehicles from intrusive magmas through restricted channels.

Metallogenetic Epochs and Provinces.—There are certain broad divisions in North America of which particular types of deposits are characteristic. These divisions in general may be assigned to different geologic epochs. Several prominent mineral belts are well known. Gold quartz veins in schistose rocks are characteristic of a zone stretching for a long distance through California and to the north and south. A series of lenticular copper deposits occurs along the foot-hills of the Sierra Nevada in California. Many silver and silver-gold deposits east of the Sierra Nevadas present points of similarity and are referred to the early Tertiary; these deposits are characteristic of a broad zone that persists for a great distance to the south through the Cordilleras. Certain areas along the Appalachians in the Eastern States and in eastern Canada, and certain areas in the Western States, are characterized by deposits of pre-Cambrian age, which likewise present many points of similarity. The distribution of the disseminated chalcocite enrichments in the southwestern desert region is well known.

A majority of western ore deposits[1] were formed in late Cretaceous or in Tertiary times, and no great epochs of dynamic stress and metamorphism have affected them.

The Distribution of Ore-Deposits in Individual Mining Districts. —The valuable ore-deposits of any mining district commonly may be referred to a single period of mineralization, or to a single set of conditions; wherever possible, these governing factors should be determined and all exploration work should be planned with regard to them. Exploration by elimination, or the demonstration that certain areas do not contain ore-deposits, rarely yields results in proportion to the money spent; all exploration should

[1] Waldemar Lindgren, *Economic Geology*, Vol. IV, p. 58.

be directed to test the hypothesis that appears most reasonable in the light of a knowledge of ore-deposits in general, and of the particular mining district in which the work is undertaken.

The principal ore-deposits of any mining district are likely to be confined to a certain series of fissures, to some particular rock mass, stratum, area of altered rocks, locus of shattering, or to the vicinity of some particular intrusive or contact; not infrequently, the payable veins of a district fall into groups of similar strike, while the barren veins or lean veins fall into other groups. The significance of these relations is apparent in directing exploration.

The Association of Ore-Deposits with Certain Rocks.—There is a persistent connection between ore-deposits and monzonitic rocks throughout the Cordilleran region; the examples of this relation include many of the most important districts. That the converse of this relation—that ore-deposits may be expected where monzonitic stocks are found—is not true, as is illustrated by the numerous monzonite masses through New Mexico that are not connected with any important mineralization.

The supposition that certain types of igneous rocks indicate the existence of ore is a fallacy that has been the cause of much fruitless exploration and loss. While a particular intrusive frequently controls ore deposition over a limited district or area, and exploration in that district is best confined to the sphere of influence of the intrusive, it does not follow that the same kind of rock elsewhere is a favorable indication of valuable deposits.

The association of tin and of tungsten with granite is well established, as is also the association between nickel, cobalt, platinum and chromium with basic rocks, commonly those that carry abundant ferro-magnesian silicates. Many other associations have been pointed out, but they appear to be persistent in restricted areas only.

The Depth to which Primary Ores Persist.—The outcrop or the exposure at any horizon of a primary ore-deposit affords a criterion of the value of that deposit to any depth above the zone of primary impoverishment; unless it can be shown that the deposit represents the root of a vein by far the greater proportion of which has been destroyed by erosion, this is likely to mean any depth attainable by mining operations.

There is no genetic reason why the values of any primary deposit should not continue to great depths. Veins that are pockety, or whose ore-shoots are short and irregular, will probably maintain these characteristics in depth, but the values, whatever their distribution, should be substantially the same at all horizons.

Secondary ores, however, being the results of surface processes, are limited to horizons near the surface; in mining geology there is no distinction of greater practical importance than that between primary and secondary ores.

The Criteria of Primary Ores.—In the investigation of any ore-shoot the first consideration is whether the ore is primary, secondary, or residual, as upon this rests all conclusions in regard to its persistency in depth. A primary ore is an ore that has undergone no change since deposition. A secondary ore is an ore formed by secondary, or surface, agencies. A residual ore is an ore that has remained after the solution and removal of associated minerals by secondary processes.

In distinguishing between primary and secondary ores, the first criterion is the presence or absence of signs of oxidation; if an ore carries traces of oxidation, secondary action must be suspected, although it may be shown that it has had no effect in the distribution of values.

In thin section under the microscope the manner of intergrowth of primary minerals is characteristic, and in this way primary and secondary ores may usually be distinguished from each other; evidence of structural intergrowth is rarely visible to the unaided eye. The presence of two generations of sulphides, the richer being in general the later, and coating the other as if precipitated upon it, is usually clear evidence of secondary enrichment. The presence of seamlets of one sulphide, especially if the richer, through another sulphide, is likewise evidence of secondary enrichment. In deposits of massive pyrite that carry chalcopyrite of undoubted primary origin, there is a tendency for the latter mineral to occur as veinlets through the pyrite. This is probably due to the greater solubility and the later crystallization of chalcopyrite over pyrite under conditions of regional metamorphism, to which most of the deposits of this kind have been subjected.

The presence of fluid inclusions in an ore is evidence of primary origin unless secondary processes are seen to have been at work; a banded structure, or crustification, is another evidence of primary origin. While there are many doubtful cases where an ore may not be assigned definitely to either primary or secondary processes, in a majority of ores the typical associations of primary minerals, their manner of intergrowth, the presence of fluid inclusions or of crustification, indicate a primary ore, while the absence of oxidation, of secondary minerals, or of secondary rearrangement of the primary minerals, indicate that surface agencies have played no part in the distribution of values; such an ore may be expected to continue in depth to the zone of primary impoverishment, which in most cases is deeper than the limits of profitable mining.

No absolute rule may be formulated for field use, but an association of the following sulphides commonly indicates a primary origin for any ore that carries no trace of oxidation or typical secondary structure: galena, zincblende or chalcopyrite with pyrite, pyrrhotite or arsenopyrite. The most common primary occurrence of gold is either as native gold or as a telluride. Probably the most common primary occurrence of silver is as argentite. Auriferous or argentiferous tetrahedrite is a common primary mineral.

The Minerals of Distinctively Primary Origin.—The minerals present in an ore frequently afford a basis upon which to judge its origin. Some minerals are distinctively primary, some distinctively secondary, others, and among them are some of the most important ore minerals, are in some instances primary and in others secondary. The presence of a mineral of secondary origin proves the action of surface agencies; a mineral that is known to be sometimes of secondary origin indicates that surface agencies may have enriched the ore under consideration, and so casts doubt upon the primary character of the ore containing it. Genetic classifications of minerals have been made by Waldemar Lindgren[1] and W. H. Emmons,[2] and their tables, which were made for another purpose, have been freely consulted in the preparation of the following lists:

Distinctively Primary Minerals

ORE MINERALS

Arsenopyrite	Pyrrhotite
Bismuthinite	Tellurides
Cobaltite	Tetradymite
Stibnite	

GANGUE MINERALS

Albite	Orthoclase
Biotite	Rhodonite
Diopside	Rutile
Fluorite	Scapolite
Garnet	Specularite
Graphite	Spinel
Hornblende	Topaz
Ilmenite	Tourmaline
Muscovite	

Minerals both Primary and Secondary in Origin

Argentite	Proustite
Bornite (usually secondary)	Pyrite
Chalcopyrite (usually primary)	Polybasite
Enargite	Sphalerite
Galena	Stephanite
Gold	Tetrahedrite
Pyrargyrite	Tennantite

[1] *Economic Geology*, Vol. II, p. 122.
[2] *Economic Geology*, Vol. III, p. 625.

Distinctively Secondary Minerals

Chalcedony	Pyrolusite
Cuprite	Chlorides
Chalcocite	Sulphates of the heavy metals.
Covellite	Carbonates of the heavy metals.
Kaolin	Phosphates of the heavy metals.
Limonite	Silicates of the heavy metals.
Opal	Arsenates of the heavy metals.

The Primary Associations of Metals.—In primary ores some metals exhibit a tendency to associate themselves with certain minerals. Among the more prominent of these primary associations are:

Gold with quartz.
Gold with pyrite.
Gold with chalcopyrite.
Silver with galena.
Silver with copper.
Silver with manganese.
Copper with pyrite.
Lead with barium.

Of these associations that of gold with chalcopyrite is probably stronger than its association with either quartz or pyrite, and the association of silver with lead is most marked.

The Accessory Minerals that Commonly Indicate a Segregation of Values.—Tetrahedrite is a guide to high silver and gold values in most deposits in which it occurs. In quartz veins, the presence of finely disseminated galena or chalcopyrite ("sulphurets"), or the presence of fluorite, are often indicative of high gold or silver values. In quartz veins that carry gold and silver it is frequent that quartz of a certain texture carries high values, while associated quartz of other textures is low grade or barren.

It is generally supposed that galena having curved crystal faces carries higher silver values than galena of cubical cleavage; it is not unusual that where the crystal faces of galena are curved the associated minerals have a similar structure; the supposed relation, therefore, is not always a reliable guide. The fact is well known that a fine grained or granular galena is likely to carry more silver than the coarsely or well-crystallized mineral. Well-crystallized pyrite is usually quite lean in copper. It seems probable that the presence of silver in galena and of copper in pyrite tend to interrupt crystallization, and it is not unusual that information as to the content of these associated or contained metals may be gained by an inspection of the casts left in the outcrop after the solution and removal of the principal sulphide.

The relations between segregations of values and accessory minerals, or varying textures, are soon learned in the study of individual deposits, and often form valuable guides in exploration.

Primary Gold Ores.—At Cripple Creek, Colorado.[1]—The characteristic feature of the ores is the occurrence of the gold in combination with tellurium, chiefly as calaverite, but partly also as the more argentiferous sylvanite, and probably to a minor extent as other gold, silver and lead tellurides. Native gold appears to be absent from the telluride ores, except where set free by oxidation. Pyrite is widely disseminated through the country rock and also occurs in small quantities in the fissures associated with tellurides. Galena and sphalerite are sparingly present in the majority of the veins; tetrahedrite and stibnite are of frequent occurrence; molybdenite in small quantities is probably always present. The tetrahedrite is usually rich in silver, and also contains gold; possibly, however, the latter is due to admixed calaverite, as the two minerals are often found in intimate intergrowth. The galena and zincblende rarely contain enough of the precious metals to form ore. Auriferous pyrite is often reported, but in the cases investigated the gold was found to be present as admixed tellurides. The usual minerals of the scanty gangue are quartz, fluorite, and dolomite.

At the Alaska-Treadwell Mine, Alaska.[2]—The ore-bodies follow a dike of albite-diorite that carries a net-work of quartz and

[1] Lindgren and Ransome, U. S. G. S., *P. P.* 54, p. 169.
[2] A. C. Spencer, U. S. G. S. *Bull.* 225, p. 39, and *Bull.* 287, p. 105.

calcite veinlets. Pyrite occurs both in the veinlets and disseminated through the rock itself. Gold occurs in association with the pyrite and also native, and a large, though variable, proportion of the value of the ore is saved by amalgamation. Visible specks of the gold are sometimes, though rarely, found. Associated minerals always present are pyrrhotite and magnetite; molybdenite is of common occurrence. Native arsenic, realgar, and orpiment have been noted. Arsenopyrite is suspected. Stibnite occurs in small amounts with the quartz. The bullion assays indicate small quantities of silver only.

AT GRASS VALLEY, CALIFORNIA.[1]—The primary ore is quartz that carries free gold in both fine and coarse particles, with from 2 per cent. to 3 per cent. of sulphides which also carry gold. Pyrite is the predominant sulphide; associated with it are galena, zincblende, chalcopyrite, and arsenopyrite. Subordinate accessory minerals are tetrahedrite and molybdenite. The quartz carries a little calcite. Fluid inclusions are abundant, and in many specimens are distributed in a manner dependent upon the distribution of the sulphides through the quartz.

Primary Copper Ores.—AT CLIFTON, ARIZONA, the primary ores are unpayable disseminations that assay about 3/10 of 1 per cent. copper; from these low-grade ores the valuable deposits have been formed by secondary enrichment. The primary ore consists of sericitized quartz-mon zonite porphyry that carries veinlets of quartz and pyrite, disseminated pyrite, and a little finely divided chalcopyrite, zincblende, and molybdenite.

AT DUCKTOWN, TENNESSEE,[2] theprimary ores consist of massive pyrrhotite containing particles and stringers of chalcopyrite and pyrite, together with minute quantities of galena and zincblende. Calcite, zoisite and quartz, and occasionally bunches of garnet occur with the ore.

IN SHASTA COUNTY, CALIFORNIA,[3] the primary ores consist of

---

[1] Waldemar Lindgren, *Seventeenth Annual Report* U. S. G. S., Pt. II, p. 124.

[2] W. H. Weed, "Copper Mines of the World," p. 349.

[3] J. S. Diller, *Bull.* 285, U. S. G. S., p. 173.

pyrite and chalcopyrite, with some zincblende and galena, associated with quartz, calcite, and barite.

AT FALUN, SWEDEN,[1] the ore is essentially a granular-crystalline mixture of pyrite and quartz, with accessory magnetite, chalcopyrite, pyrrhotite, zincblende, and in rare cases, galena.

AT BISBEE, ARIZONA,[2] the primary ore consists of pyrite containing variable amounts of chalcopyrite and a little sphalerite associated with calcite, amphibole, pyroxene, garnet, chlorite, quartz, and vesuvianite; the metamorphic silicates are usually so finely divided as to be indistinguishable by the unaided eye. The primary ore in general is unpayable.

Contact Ores.—Typical contact ores carry pyrite, chalcopyrite, bornite, pyrrhotite, specularite, and magnetite, with commonly lesser amounts of galena and zincblende, associated with garnet, wollastonite, epidote, amphibole, pyroxene, vesuvianite, quartz and calcite. The gold and silver content is usually low. The distinctive association of minerals in contact ores is that of primary oxides with sulphides.[3]

AT SAN PEDRO, NEW MEXICO,[4] the contact ores consist of massive garnet replacing limestone and carrying chalcopyrite, specularite, epidote, vesuvianite, wollastonite, quartz and calcite.

Primary Silver Ores.—AT LAKE CITY,[5] COLORADO, the primary ore minerals are galena, tetrahedrite, chalcopyrite, sphalerite and pyrite, associated with quartz, rhodonite, rhodochrosite, and barite. The silver is contained in the galena to the extent of 22 to 30 oz. per ton, and in the tetrahedrite, which is probably related to freibergite, in much larger quantity.

AT THE GRANITE-BIMETALLIC MINE, MONTANA,[6] the primary ore consists of pyrite, arsenopyrite, tetrahedrite, and tennantite, with lesser quantities of galena and zincblende, in a gangue of

[1] "Nature of Ore Deposits," Beck-Weed, p. 460.

[2] F. L. Ransome, *P. P.* 21, U. S. G. S.

[3] Waldemar Lindgren.

[4] M. B. Yung and R. S. McCaffery, Trans. A. I. M. E., XXXIII, p. 355.

[5] J. D. Irving, *Bull.* 260, U. S. G. S., p. 81.

[6] W. H. Emmons, *Bull.* 315, U. S. G. S., p. 39.

quartz and rhodochrosite. Sparingly scattered through this ore are found small specks of pyrargyrite, proustite, and, rarely, realgar and orpiment. This ore carries from 20 to 30 oz. silver and from $1.50 to $3.00 in gold.

AT GEORGETOWN, COLORADO,[1] the primary ore consists of argentiferous galena and zincblende, with cupriferous pyrite, and chalcopyrite, in a gangue of quartz, and siderite, rhodochrosite, dolomite and calcite, in varying proportions; fluorite is present, but is rare.

AT TONOPAH, NEVADA,[2] the primary ores consist of argentite, polybasite, stephanite, and gold in a still undetermined form, associated with chalcopyrite, pyrite, and subordinate galena and zincblende in a gangue of quartz, adularia, sericite and carbonates.

AT PARK CITY, UTAH,[3] the primary ores consist of galena tetrahedrite, pyrite, chalcopyrite and zincblende in a siliceous gangue that carries a little barite and fluorite.

AT PACHUCA, MEXICO,[4] the primary ores where the veins become impoverished in depth consist of galena, zincblende, and pyrite in a gangue of quartz and rhodonite.

AT GUANAJUATO, MEXICO,[5] the primary ores consist of tetrahedrite with zincblende and pyrite in a gangue of quartz, calcite, rhodonite, and fluorite.

Primary Lead Ores.—AT LEADVILLE, COLORADO,[6] the primary ores consist of limestone replaced by quartz, pyrite, galena and zincblende, carrying a small quantity of silver sulphide. The ore contains about 1 per cent. manganese, in what form is not stated, as rhodonite and rhodochrosite are absent.

IN THE COEUR D'ALENE DISTRICT, IDAHO,[7] the primary ore consists of galena, pyrite, pyrrhotite, chalcopyrite, sphalerite,

[1] J. E. Spurr, *P. P.* 63, U. S. G. S., p. 136.
[2] J. E. Spurr, *P. P.* 42, U. S. G. S., p. 22.
[3] J. M. Boutwell, *Bull.* 225, U. S. G. S., p. 147.
[4] Srs. Aguilera and Ordonez, *Trans.* A. I. M. E., XXXII, p. 224.
[5] Beck-Weed, "Nature of Ore Deposits," p. 265.
[6] S. F. Emmons, *Mono. XII*, U. S. G. S., p. 32.
[7] F. L. Ransome, *P. P.* 62, U. S. G. S., p. 107.

and a little tetrahedrite and stibnite. Siderite and a little quartz form the gangue. Tetrahedrite is commonly accompanied by high values in silver.

Primary Zinc Ores.—The only primary ore of zinc of importance is zincblende; in characteristic occurrences it is associated with pyrite, and occasionally with galena and chalcopyrite.

The Depth of Primary Ore Deposition.—An important factor in the consideration of the probable behavior of a primary deposit in advance of exploration is the depth at which it formed. If it can be shown that a deposit was formed at the present surface, then it is evident that the type of ore prominent at the surface cannot be expected to continue in depth, as the conditions under which the ore was deposited—namely, surface conditions—were absent.

A deposit formed at relatively shallow depth is likely to grow more regular with deeper exploration; stringers and branch veins commonly consolidate in a single lode, or lodes, of relatively greater uniformity of dip, strike and thickness as compared with the more scattered units near the surface. This advantage of regularity is likely to be offset by a decrease in size in the deeper parts of the deposit.

A deposit of deep-seated origin, however, owes its discovery to a deep erosion, which has presumably removed the upper and irregular portions of the deposit, and exposed the deeper zone of relatively greater regularity. By regularity is meant the relative regularity of different parts of the same deposit, and not the absolute regularity of a particular deposit as compared with deposits in general. Some primary deposits are probably quite irregular at all depths. A deposit of deep-seated origin whose values have not been redistributed by surface agencies is likely to be persistent in depth down to the zone of primary impoverishment, and any section of such a deposit may be considered as indicative of its character at other horizons, in the absence of the factors that cause the localization of values into ore-shoots.

The deepest parts of ore-deposits, usually referred to as the roots of the veins, are not so readily recognizable as the zones

6

that have been referred to, and the data in regard to them are less satisfactory.

That increasing depths should give rise to transition types is to be expected, and it is not possible, therefore, to divide deposits into sharply delimited classes on the basis of their depths of formation.

Deposits Formed at the Surface.[1]—Primary mineral deposits formed at the surface by hot waters are rarely of economic importance; their primary condition is commonly obscured by the action of surface agencies. The sinters characteristic of surface-formed deposits are commonly made up of silica, as opal or chalcedony, and earthy carbonates. Calcite, fluorite, celestite, barite, and many other gangue minerals may also develop in crystallized form. Stibnite, pyrite, marcasite and cinnabar are known in crystallized form, and many other sulphides have been detected chemically in such deposits. Surface waters containing atmospheric oxygen are likely to have altered these deposits greatly, and limonite, hydrous oxides, carbonates, and sulphates of the heavy metals predominate among the ore minerals, and kaolin, allophane and chloropal among the gangue minerals. The surficial formation of these deposits is commonly indicated by their structure.

Veins Formed near the Surface.—In typical examples these veins cut through beds of relatively recent volcanic rocks, and their depths at the time of vein formation may usually be determined with fair accuracy. Structural features indicating a formation at shallow depth are:[2] a greater number and width of the fissures near the surface; a branching of the upper parts of fissures; and fissures of changing dip, of which the deeper part is likely to have the flatter dip.

Metasomatic alteration of wall rocks is likely to extend to greater distances from veins formed near the surface than from deeper veins. In rocks of medium acidity a strong sericitization

[1] This and succeeding paragraphs are taken from Mr. Waldemar Lindgren's articles in *Economic Geology*, Vol. II, p. 460, and *Economic Geology*, Vol. I, p. 34.

[2] Waldermar Lindgren, *P. P.* 54, U. S. G. S., p. 167.

is common immediately along the veins, and extensive pyritization is also frequent; the altering solutions have a tendency to spread from the veins through a large area of adjoining rocks, owing to more extensive fissuring near the surface, where, robbed of their most active ingredients, they effect a propylitic alteration over large areas. In very basic rocks this propylitization extends close up to the veins, where sericitization is likely to take its place; in large, irregularly altered areas, especially in siliceous rocks like rhyolite, extensive silicification is common.

In these deposits gold and silver prevail, and as compared with deep-seated veins of quartz gangue, silver is relatively more abundant, and free gold is commonly present in a more finely divided form; pyrite, zincblende, chalcopyrite, arsenopyrite, argentite, tellurides and stibnite are the prevailing ore minerals; among gangue minerals, quartz is most abundant, but it is often accompanied by chalcedony or opal; calcite and dolomite are moderately abundant in the vein filling; siderite occurs more rarely; barite and fluorite predominate locally. Magnetite and specularite, as well as all the silicates belonging to the greater depths of ore deposition, are absent. In these veins the filling of open spaces is an important process.

Veins of Deep-seated Origin.—These veins are divided by Mr. Lindgren into four classes:

(a) Contact deposits, which are discussed elsewhere.

(b) Cassiterite veins. The characteristic minerals of this type are cassiterite, pyrite, arsenopyrite, specularite, quartz, tourmaline, topaz, lepidolite, muscovite, apatite, fluorite, and wolframite, with subordinate calcite and siderite. These veins are commonly poor in gold and silver, and the metasomatic alteration along their walls is likely to be intense.

(c) Apatite veins. The characteristic minerals of this type are apatite and other phosphates, scapolite, diopside, hornblende, biotite, specularite and pyrrhotite; strong metasomatic action is usual along the walls of these veins, and the introduction of chlorine and fluorine is characteristic; these veins are commonly poor in gold and silver.

(d) Deep-seated gold and silver veins. The characteristic minerals of this type are gold, pyrite, pyrrhotite, galena, zinc-blende, magnetite, specularite, ilmenite, quartz, biotite, tourmaline, garnet, hornblende, chlorite, apatite, spinel, and epidote; calcite is present in small amounts; the replacement of the country rock is usually strongly marked; amphibolites and micaceous schists are replaced by tourmaline, garnet, a green biotite and epidote; soda-lime feldspars are unstable under the influence of vein forming solutions and alkali feldspars usually do not form. These veins commonly occur in, or close to, granite intrusives in schists.

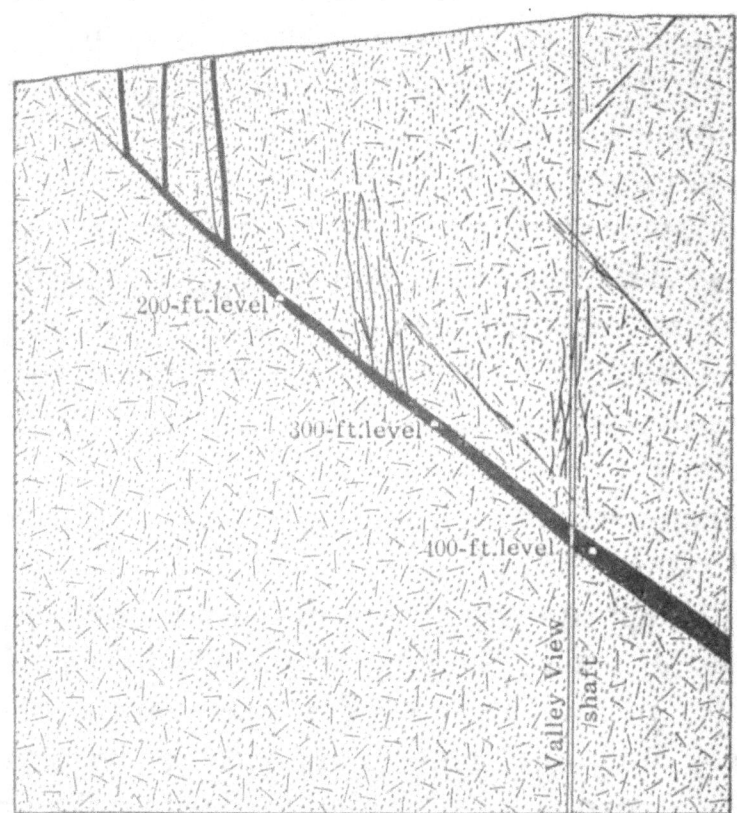

Fig. 41.—Section of the Valley View vein, Tonopah, Nevada, showing upward branching. *After Spurr.*

As a rule, the walls of deep-seated veins are not altered through so great distances as is common with veins formed at shallow depths, and their vein fillings frequently bear evidence of having been subjected to the stresses of dynamometamorphism. The time of formation of the deep-seated deposits is likely to be remote as compared with deposits formed at shallow depths.

Relative Susceptibility of Hanging- and Foot-walls to Mineralization.—In many veins the hanging-wall has been subject to brecciation and mineralization to a far greater extent than the foot-wall, and hydrothermal alteration more frequently extends into it than into the foot-wall. This is probably the result of the superior resistance of the foot-wall to fracturing as compared with the hanging-wall. The hanging-wall rocks readily adjust themselves to stress through fracturing, while the foot-wall, under as great, or greater, stress, remains massive, owing to the reinforcement by underlying rock masses.

CHAPTER V

TYPES OF PRIMARY ORE-DEPOSITS

The classification here used is one of convenience only; it is not intended to include all known types of ore-deposits. The characteristic features of the several well-marked types of primary mineralizations are described without reference to the ultimate source of their metals. The minerals of many deposits of obscure origin were probably introduced through fissures that are now healed, or are due to the migration of metals from such deposits. A discussion of the genesis of these deposits is not justified by the present state of knowledge of economic geology.

Magmatic Segregations.—During the solidification of magmas under conditions that do not permit the escape of their metallic

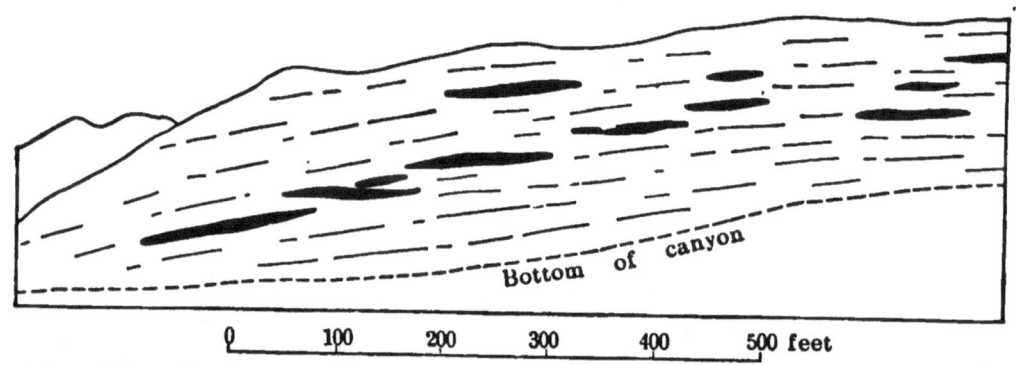

Fig. 42.—Sketch showing the Drinkwater zone of quartz lenses in alaskite, Silver Peak, Nevada; the lenses are magmatic segregations. *After Spurr.*

content, there is a tendency for like particles to form segregations probably through mass action, or the mutual attraction of like particles. Magmatic segregations are commonly made up of compounds of relatively low mobility.

Basic rocks appear to be the most suitable for the formation of segregations, the characteristic minerals of which are magnetite, ilmenite, chromite, pyrrhotite, pyrite, and pentlandite. The gangue minerals are those of the containing rock.[1] Acid, or

[1] Waldemar Lindgren, *Economic Geology*, Vol. II, p. 110.

quartzose segregations, however, are known. Gold and platinum have been established as constituent minerals of rock masses, chiefly as sparse disseminations, unpayable in themselves, but important as the sources of placer deposits.

Magmatic segregations are rarely of economic importance as ores, and may usually be recognized in thin section under the microscope by the manner of intergrowth of the ore with the rock minerals. It is not unusual to find that rocks containing such deposits have become further differentiated into two or more parts of different mineralogical composition. Clearly established examples of magmatic segregations are more rare than was at one time supposed. The most generally accepted type is that of the titaniferous magnetite deposits.

IN THE ADIRONDACK MOUNTAINS, NEW YORK,[1] deposits of titaniferous magnetite are associated with basic intrusives in such a way as to indicate that the ore minerals were segregated from the containing rock as it cooled. No ore was formed along the contacts of these intrusives with the enclosing rocks, and the ore grades into the containing rock through a transition zone that shows a gradual change in the quantities but not in the kinds of the constituent minerals.

AT SILVER PEAK, NEVADA,[2] auriferous quartz appears to have segregated from enclosing alaskite, which is a phase of the accompanying granite; the ore bears the same relation to the alaskite that the latter bears to the granite. The alaskite is a granite without biotite; the quartz is an alaskite without feldspar. The segregated quartz is in this case of the same age and generation as the granules of quartz that make up a large proportion of the alaskite and granite.

Contact Deposits.—Contact deposits are deposits formed along the contacts between intrusives and their enclosing rocks, or in these rocks in immediate proximity to the intrusives. They are the result of direct emanation of mineral-bearing solutions from

[1] J. F. Kemp, *Nineteenth Annual Report*, U. S. G. S., III, p. 392.
[2] J. E. Spurr, *P. P.* 55., U. S. G. S., p. 87.

the intruding magma.   Contact deposits are usually limited to
rocks that exert strong precipitative action.   The characteristic
minerals of contact deposits are:[1] specularite, magnetite, bornite,
chalcopyrite, pyrite, pyrrhotite, and more rarely, galena and
zincblende, associated with garnet, wollastonite, epidote, ilvaite,
amphibole, pyroxene, zoisite, vesuvianite, quartz and calcite,
and rarely, fluorite and barite.   The unique feature is the association of the oxides of iron with sulphides.   The sulphides frequently carry gold and silver, but usually in small quantities
only.

Contact deposits are rarely of commercial importance, although
there are certain notable exceptions.   Many deposits formerly
included in this class are now more correctly considered as
replacement deposits.

Contact deposits are usually quite irregular in form, due largely
to the common irregularity of the igneous intrusion, and when
the ore is lost, it is generally recovered with difficulty; while
there is apparently no genetic reason why contact deposits,
which are of deep-seated origin, should not be persistent, it is
well known that few contact deposits are mined at more than
shallow depths.

In most cases certain beds of the intruded rocks exert a greater
precipitative action, or are more easily replaced, than other beds
of the series, and contact minerals develop much more abundantly
in them than elsewhere.   It is necessary in the examination of
contact deposits to trace these favorable beds and to determine
their probable thickness beneath the better exposures; not infrequently, the important mineralization is confined to one or more
such beds, to which, therefore, exploration should be limited,
and below which the mineralization should not be expected to
extend.

Most contact minerals are resistant to weathering, and erosion
is often halted at the horizon where the contact metamorphic
silicates reach their largest development.   Large outcrops do
not, therefore, afford a reliable criterion of the extent of such

[1] Waldemar Lindgren, *Trans.*, A. I. M. E., Vol. XXXI, p. 227.

deposits in depth. A large development of contact metamorphic silicates does not necessarily indicate the existence of valuable ore any more than a barren quartz vein indicates the presence of gold. Whether or not such an outcrop contained sulphides before oxidation may usually be determined by a search for and examination of the casts left in the resistant silicates upon the solution of contained sulphides; this feature is taken up in the chapter on outcrops. The ores of contact deposits, unless of smelting grade, offer serious metallurgical difficulties, owing to the high specific gravity of their gangue minerals.

The phenomena of contact metamorphism are considered in the chapter on Primary Alterations of Wall Rocks.

AT MORENCI, ARIZONA,[1] important contact deposits occur in limestones and shales along intrusions of quartz-monzonite porphyry. Wherever the porphyry came in contact with granite or quartzite, little alteration is observed, but wherever the porphyry meets the limestones or shales of the Paleozoic series extensive contact metamorphism has taken place. The whole Paleozoic series is affected, but more particularly the pure limestone of the lower Carboniferous, which, for a distance of several hundred feet from the contact, has been converted into an almost solid mass of garnet. The shales have suffered less from this metamorphism, but near the porphyry are likely to contain epidote and other accessions. Wherever alteration by surface agencies has not masked the phenomena, magnetite, pyrite, chalcopyrite, molybdenite, specularite and zincblende accompany in various proportions the contact-metamorphic minerals, which here comprise garnet, epidote, diopside, tremolite and quartz. In form, the deposits in limestone are irregular, but in many cases they assume a tabular shape, due to the accumulation of the minerals along certain favorable planes of stratification, or along the walls of dikes. These contact deposits differ from most examples in the absence of wollastonite and vesuvianite, and in the metamorphism of the shales; instead of the knotty schist or hornfels usually produced from shales there is found at Morenci a greenish horn-

[1] Waldemar Lindgren, *P. P.* 43, U. S. G. S., p. 19.

fels, with much amphibole (tremolite), epidote, pyrite and magnetite.

Pegmatitic Deposits.—The characteristic minerals of pegmatitic deposits[1] are magnetite, bornite, arsenopyrite, molybdenite, cassiterite and wolframite, associated with quartz, muscovite, alkali feldspars, tourmaline, apatite, fluorite, spodumene, and more rarely, hornblende and soda-lime feldspars. These deposits, which are of deep-seated origin, contain little gold and silver and except where associated with stockworks and cassiterite-bearing impregnations they are irregular in value and of slight economic importance, except for mica and minerals of the rare earths. Pegmatite dikes carrying wolframite are known at many places in the western United States; pockets of rich ore are occasionally found in them, but these deposits apparently have not repaid exploration.

Fahlbands.—Beds of schist that have been impregnated with sulphides and subjected to dynamo-regional metamorphism are known as fahlbands.[2] The sulphides occur disseminated through the schist intergrown with the principal minerals of the rock, and also along the planes of schistosity; the association is such as to render obscure their origin and mode of introduction. These deposits, which are rarely of economic importance, are apparently confined to areas of pre-Cambrian rocks. Fahlbands frequently persist over long distances, but it is rare that their mineralization is sufficiently concentrated to form ore. In certain European localities, they have exerted an important influence upon the segregation of values in veins that cross them.

IN THE UPPER PECOS DISTRICT, NEW MEXICO,[3] an amphibolite, probably produced by regional metamorphism from a dioritic or diabasic rock, carries disseminated chalcopyrite and zincblende that contain a little gold and silver, associated with a green biotite, tourmaline, and veinlets of dark quartz, and intergrown with the principal minerals of the containing rock.

---

[1] Waldemar Lindgren, *Economic Geology*, Vol. II, p. 111.
[2] J. F. Kemp, "Ore Deposits," p. 73.
[3] Lindgren, Graton and Gordon, *P. P.* 68, U. S. G. S., p. 50.

Regionally Metamorphosed Ore-Deposits.—Deposits that were formed in remote geological ages are likely to have been deeply buried, and to have undergone rearrangement under the stresses of dynamo-regional metamorphism. The recrystallization result-

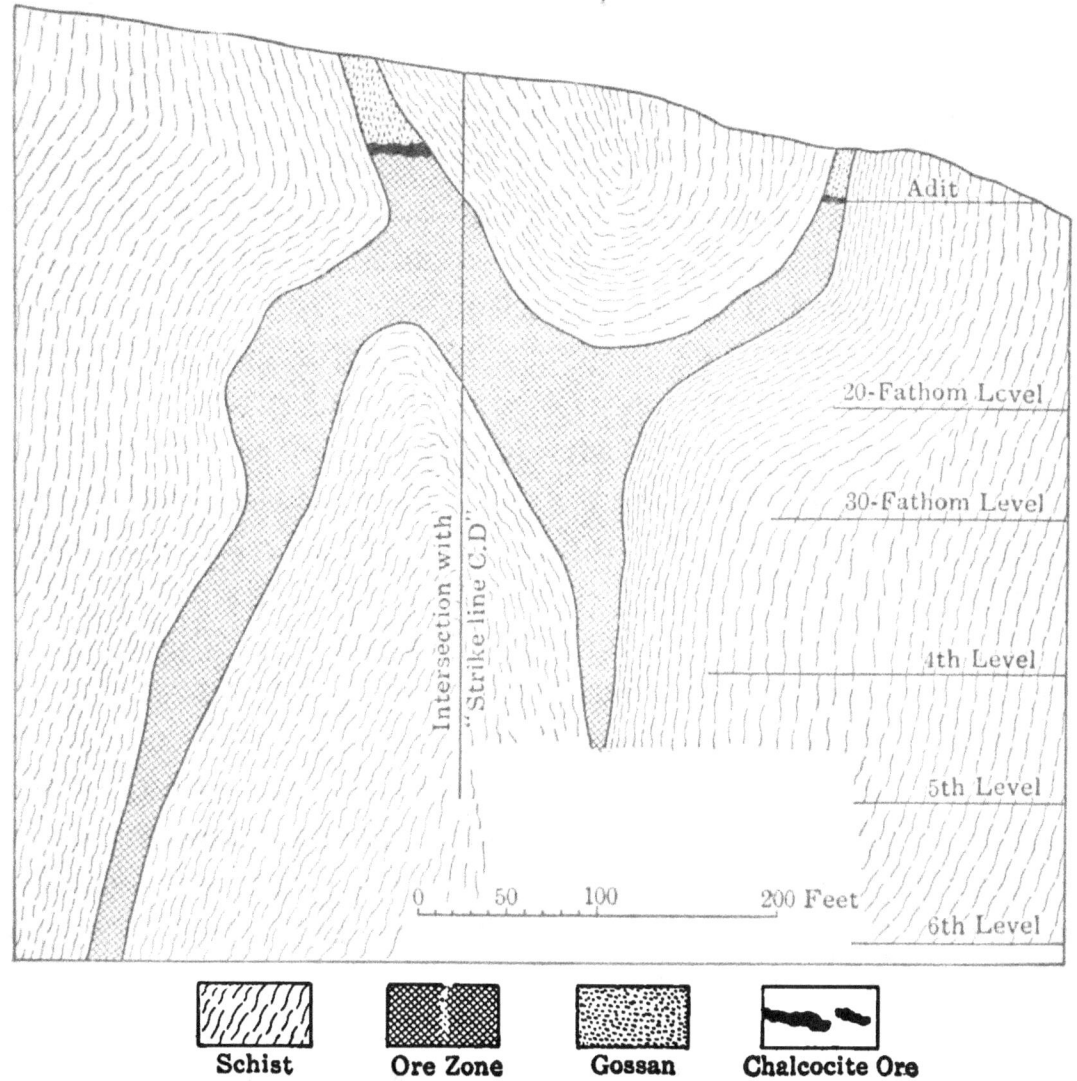

Schist   Ore Zone   Gossan   Chalcocite Ore

Fig. 43.—Section of the ore-deposit in the Mary mine, Ducktown, Tennessee, showing a regionally metamorphosed ore-deposit. *After Emmons and Laney.*

ing from these processes may so completely change the enclosing rocks to crystalline schists as to hide completely their original character, it being difficult to distinguish schists that result from the metamorphism of sedimentary beds from those that were originally igneous rocks. Under these conditions of pressure, heat and flowage the small amount of moisture contained in the rocks appears sufficient to accomplish a complete rearrangement of the minerals. Ore-deposits contained in these rocks have undergone complete transformation, and their original relationships have been obscured to such an extent that their origin is not determinable; they are, therefore, best described as a separate type. Rearrangement under these conditions brings about a segregation of ore minerals into homogeneous bodies, and may be considered a process of primary concentration, the ores resulting from rearrangement without accession of material from without. Minerals that are stable under the conditions of dynamo-regional metamorphism and which therefore are characteristic of these deposits are:[1] pyrite, chalcopyrite, pyrrhotite, magnetite, quartz, muscovite, biotite, epidote, hornblende and albite.

Regionally metamorphosed deposits usually occur parallel to the schistosity of the enclosing rock, or cross it at low angles, in accord with the law that tabular bodies under stress and flowage, tend to orient themselves in the direction of least pressure.[2] The relations of contact deposits to igneous rocks are likely to have been destroyed by the metamorphism, and the fissures and channels that produced replacement deposits are likely to have been healed. The usual guides in exploration for these deposits, therefore, are obliterated. The most marked structural features of regionally metamorphosed ore-deposits are their lenticular form, and the frequent distribution of such lenses in an overlapping series. The occurrence of these primary ore-bodies is discussed under "Lenticular Ore-Shoots" in a succeeding chapter. The individual lenses of deposits of this type are occasionally of large size, but not infrequently the mineralization is confined to a single lenticular mass, and expectations of future ore are not justified beyond the probable content of the lenses already exposed. These deposits, being thickest at their centers and narrowing to extinction toward their peripheries, often permit approximate estimates of their content to be made in advance of

[1] Waldemar Lindgren, *Economic Geology*, Vol. II, p. 127.
[2] W. H. Emmons, *Economic Geology*, Vol. IV, p. 776.

exploration. The section presented by the surface is likely to be a fair criterion of the distribution of deposits of this type in depth. In exploration for further deposits cross-cuts should be driven near the extremities of the known lenses to expose possible overlapping bodies; in most cases little other exploration is justified.

THE VERMONT COPPER BELT[1] contains three districts—Corinth, Copperfield and South Strafford. The deposits occur along a due north-south line, which corresponds to the general direction of the schistosity of the rocks. The rocks are micaceous schists and gneisses, formed from sandstones and shales by regional metamorphism. The original bedding, though obscure, is occasionally in evidence, and does not correspond with the foliation. Intrusions of granite are common in the region, but do not occur in the immediate vicinity of the ore-deposits. The ore bodies are lenticular masses which simulate bedded deposits, since they appear to conform to the banding of the enclosing schists. At each locality only one workable lens has been found outcropping; in the deep mines the outcropping lens wedges out in depth, but is found to overlap the tapered upper end of another lens in the foot-wall. The deposits have no gossan cap, sulphides appearing at the surface. The ores consist of massive pyrrhotite, chalcopyrite, pyrite, and a little sphalerite mixed with variable quantities of quartz and actinolite; in the leaner ores, garnet and biotite are present.

Deposits Due to the Filling of Open Spaces.—The filling of open fissures or preëxisting cavities in rocks is a process of great importance in ore deposition; while many veins are due to this process alone, a majority of mineral veins are probably the result of both replacement and the filling of open spaces. Open spaces are caused by irregular fissuring accompanied by a moderate movement, sufficient to bring projection opposite projection and concavity opposite concavity, and thus cause pinches and swells in the fissure and resulting vein.

It is clear that a uniform open space of large extent cannot

[1] W. H. Weed, *Bull.* 225, U. S. G. S., p. 199.

remain open along a fissure for any length of time, but must soon be closed by the pressure of overlying rocks. Fissures, therefore, that remained open for sufficient lengths of time to become filled and mineralized are commonly irregular in cross-section, the open stretches being separated by tight portions where the wall rocks came together and formed the buttresses that permitted intervening portions to remain open. It is a common fallacy that filled fissures are characterized by uniformity of direction and thickness; in most veins, which are the result of both filling and replacement, the latter process tends to counteract and to obscure the irregularities of the orignial open spaces.

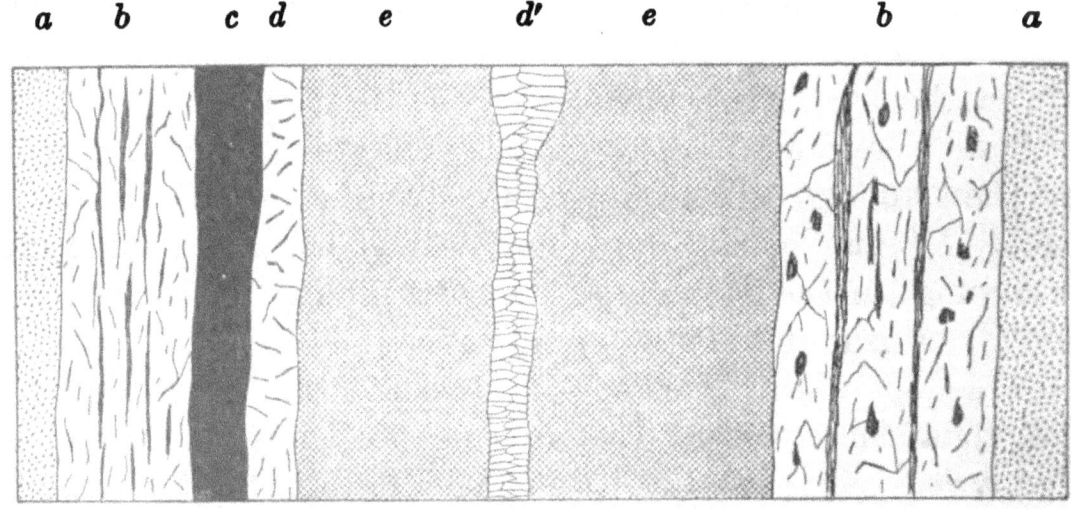

Fig. 44.—Section of a crustified vein near the London shaft, Silverton, Colorado. *a*, Country rock; *b*, quartz and cholcopyrite; *c*, tetrahedrite; *d–d'*, quartz; *e*, galena. *After Ransome.*

Enlargement by solution is thought by many to play a large part in the formation of cavities that are subsequently filled with ore. Except in limestone; perhaps, such solution takes place contemporaneously with and constitutes a part of the process of replacement.

Open fissures and cavities in rocks are more abundant near the surface than in depth, because of the less pressure of overlying rocks that tends to close them. While a majority of deep-seated veins are replacement veins, open cavities frequently persist to great depths, especially in veins that pinch and swell as before described.

According to C. R. Van Hise[1] the zone of flowage from pressure of overlying rocks, in which no cavities can exist, is reached at 1625 ft. in soft shales and at 32,500 ft. in firm granites.

Fissuring accompanied by sufficient movement to form open spaces in one rock, may produce a tight fissure in a more plastic rock at the same depth, and veins that are the result of the filling of open spaces may be expected to vary in width according to the character of the rock traversed; a vein that cuts a series of different beds may be expected to vary markedly in passing through them.

If the valuable mineral in a filled fissure was among the earliest deposited, it will be found near the walls, and may be expected to persist longitudinally along the vein; if it was among the last deposited, it is likely to be limited to the spaces at the center that remained open at the time of its deposition, and, therefore, to be more markedly confined to shoots.

The usual criterion of a filled fissure is the arrangement of the filling in crusts or bands parallel to the walls, similar bands occupying the same relative positions on either side of the center line, which is frequently marked by interlocking combs of crystals. The bands nearest the walls represent the minerals first deposited, and the central part the latest deposition. In cases where the vein after being filled has been reopened, the additional crusts deposited will not be symmetrical with respect to the original arrangement, and occasionally veins are seen that in this way exhibit a record of repeated opening and filling.

While in filled deposits evidence of crustification is often visible, not infrequently the filling is homogeneous or quite irregular, and the method of vein filling is not apparent, unless disclosed by the examination of thin sections under the microscope. The presence of radiating clusters of crystals is indicative of deposition in an open space, as is also a lack of alteration of included fragments of the wall rocks. Not infrequently,

[1] *Sixteenth Annual Report* U. S. G. S., I, p. 312.

replacement veins exhibit a banded structure; this is the result of a thin sheeting of the country rock, the lines of which are repeated and preserved in the arrangement of the vein minerals, the replacement having taken place along the sheeting planes first, and later, under different conditions, having penetrated the intervening slabs. A reopening of a fissure during or after mineralization may give rise to a structure in close imitation of crustification.

The lines of demarkation between ore and wall rock are commonly well-defined in filled deposits, while in replacement deposits the ore is likely to merge gradually into the walls. That the action of the solutions was confined to the open spaces in filled fissures is not invariably the rule; frequently the ore is found along a well-defined wall, beyond which the ore minerals do not penetrate, but the rock bordening the vein is altered, evidently by the action of the vein-forming solutions. It is probable in such cases that the walls of the vein acted as dialyzers, restraining the passage of certain elements and compounds, which precipitated within the vein, but permitting the passage of the solutions that altered the adjacent rocks.

At Pinos Altos, New Mexico,[1] the process of open fissure filling is well illustrated in the Pacific vein. Five distinct bands may be counted. Proceeding from each wall inward to the center, each of these bands has an almost perfect counterpart on the opposite side of the vein. The first band, that next to the wall, contains quartz and pyrite; its inner edge is outlined by the crystalline terminations of quartz prisms, a beautiful example of comb structure. The succeeding band is composed of zinc-blende and chalcopyrite; the chalcoprite grows more abundant toward the inner edge and, in fact, forms two subsidiary bands separated by a thin band of sphalerite. The next layer, a thin band, contains quartz and chalcopyrite. It is followed by a narrow band of sphalerite, which in turn is followed by a thicker band of quartz that contains fine grains of disseminated chalcopyrite, and locally fails to join with its corresponding band on

[1] Sidney Paige, *Bull.* 470, U. S. G. S., p. 114.

the opposite side, leaving an open crystalline cavity at the center.

On one wall of the vein is a narrow secondary vein, evidently a reopened fissure. Its walls are outlined by narrow bands of quartz (with a little chalcopyrite and galena), between which is a pinkish cream-colored mass of iron and magnesia carbon-

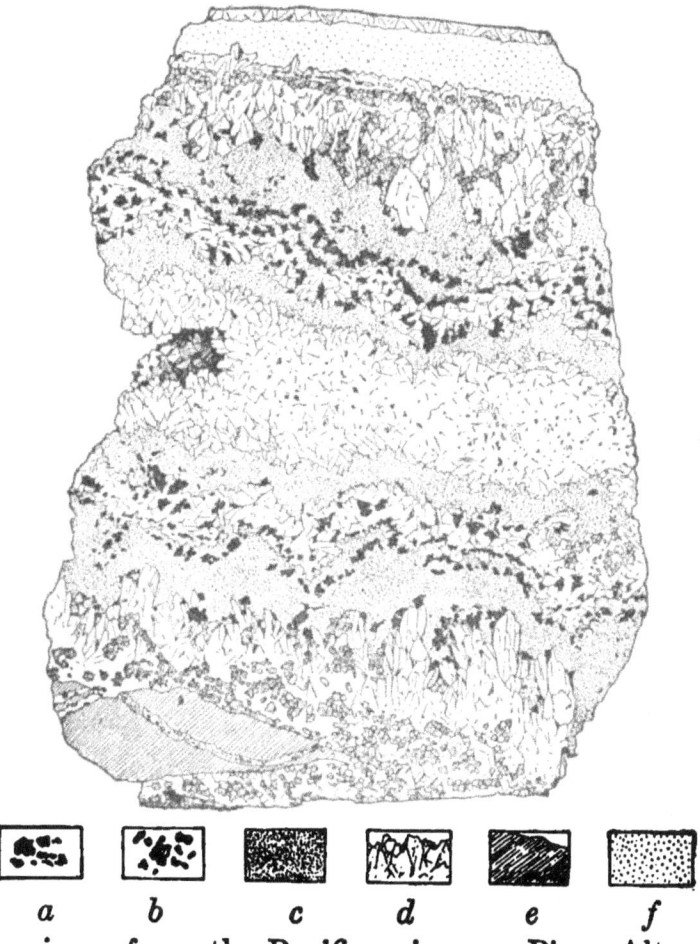

Fig. 45.—Specimen from the Pacific vein near Pinos Altos, New Mexico, showing crustification. *a*, Chalcopyrite; *b*, pyrite; *c*, zincblende; *d*, quartz; *e*, sericitized porphyry; *f*, carbonates, quartz, and iron oxide. *After Paige.*

ates, and quartz. The narrow quartz bands forming the walls of this little vein have locally been broken, and pieces of the wall now lie at varying angles across the vein, embedded in the vein filling.

On the opposite wall of the main vein a fragment of country rock is included in and surrounded by the vein material of the large vein.

7

From this data the history of the mineralization of this particular vein may be deduced. A fracture was formed in the country rock and filled by solutions carrying zinc and iron sulphides. Fracturing continued and cross fissures on a small scale were opened. The forces, of whose presence this first fracturing was a preliminary, finally succeeded in producing an open fracture measured by the width of the vein described, and solutions carrying silica, iron sulphide, a trace of zinc sulphide, and lead, circulated through the open spaces thus afforded. Along both walls quartz and pyrite were precipitated simultaneously, and continued to be precipitated, apparently, until solutions ceased to circulate, or ceased to carry sulphur, iron and silica, for the boundary between the first band and the succeeding one is sharp both in demarkation and in mineral content. When mineralizing waters next flowed past the walls, zincblende and chalcopyrite were deposited, and it is evident that, although copper, sulphur and iron were present during the remainder of the history of the vein, though growing markedly less toward the end, the zincblende and silica content fluctuated, first a layer of one and then of the other being precipitated. Parts of the vein along the center were probably never completely filled, not because there was a lack of material, but because deposition fortuitously isolated geode-like open spaces within which the circulation ceased. The small vein at the edge of the large one points to a recurrence of fracturing, and the advent of carbonated waters, carrying silica also; it marks a distinct change in the solutions, with the cessation of which the mineralization closed.

Replacement Veins.—A majority of veins are in part, at least, the result of the replacement of their walls by mineralizing solutions, and in many cases the process of fissure filling was probably so subordinate as to be practically negligable; the original fissures of replacement veins, which were probably narrow, acted chiefly as channels for the passage of the replacing and mineralizing solutions. While in a filled fissure the width of the vein represents the width of the cavity filled, and the line of demarkation between ore and wall rock is sharp, in a replacement vein the width bears no relation to the size of the original fissure, and the line of demarkation between ore and wall rock is commonly ill-defined.

Replacement veins vary greatly in size according to the ease of solubility or replacement of the rock traversed, and their values are likely to be localized by the chemical precipitative action of the different rocks traversed. Replacement veins are commonly accompanied by metasomatic alteration of their walls as will be described in a succeeding chapter; fragments of the

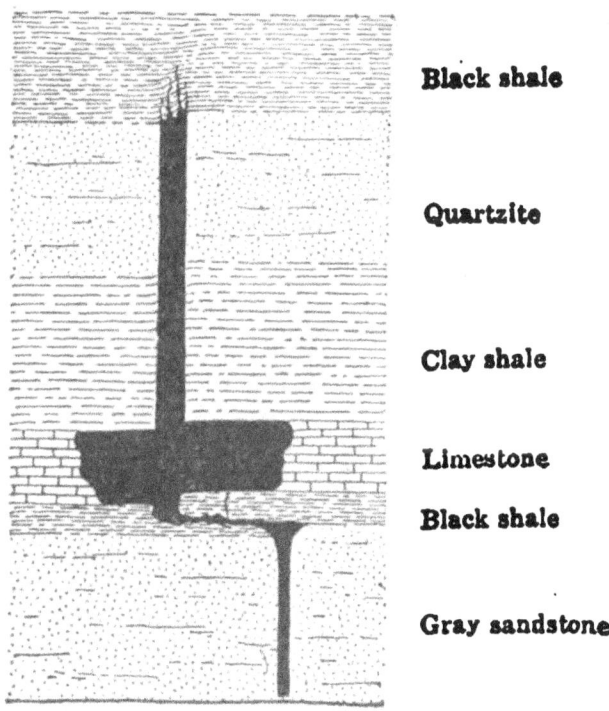

Black shale

Quartzite

Clay shale

Limestone

Black shale

Gray sandstone

FIG. 46.—Type of silver bearing vein modified by replacement, Ouray, Colorado, showing the varying size in different rocks. *After Irving.*

wall rocks included in replacement veins are likely to be partly or completely replaced by ore; the original outlines of such fragments are usually preserved in the arrangement of the replacing minerals, which may enclose cores of fresh or partly altered rock, and thus conclusively prove the process of replacement.

Replacement Deposits.—Metasomatic replacement of the country rock by mineralizing solutions is a process of great importance in ore deposition, and there are probably few epigenetic deposits in the formation of which it has not played some part. In a classification of ore-deposits for the purpose of study and practical correlation it appears best to discuss under this head those replacement deposits that are relatively homogeneous, taking up the other types according to their most prominent characteristics.

Replacement deposits are most common in the more easily soluble and replaceable rocks, among which limestone is most prominent. That the precipitative action of the replaced rock

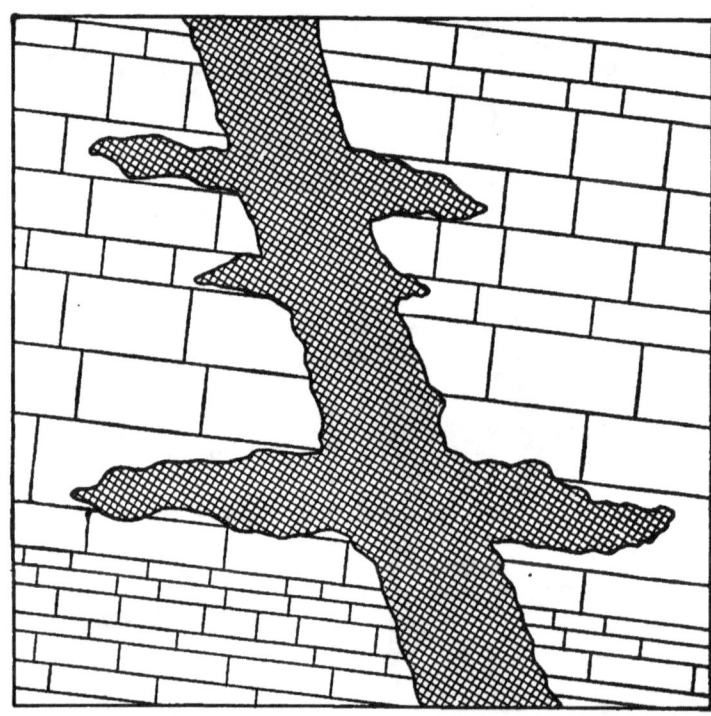

Fig. 47.—Replacement of limestone by copper ore, Bingham, Utah, showing the greater replacement of selected beds. *After Boutwell.*

is not the controlling feature of the process is shown by the occurrence of large replacement deposits in quartzite, shales, schists and other rocks as well as in limestones. The relative replaceability of a rock varies, of course, with the chemical composition of the mineralizing solution, certain solutions attacking limestone with greatest activity, while others replace quartzite with equal ease; in a given deposit, however, a rock that has been extensively replaced must be considered the ore-bearing rock, as it is unlikely that a different rock will have yielded equally to the attack of the same mineralizing solutions; important deposits in a bed of limestone, for example, may not be expected to continue into underlying quartzite or granite. It often happens that one bed only of a series of apparently similar rocks has offered a favorable horizon for ore deposition, other beds

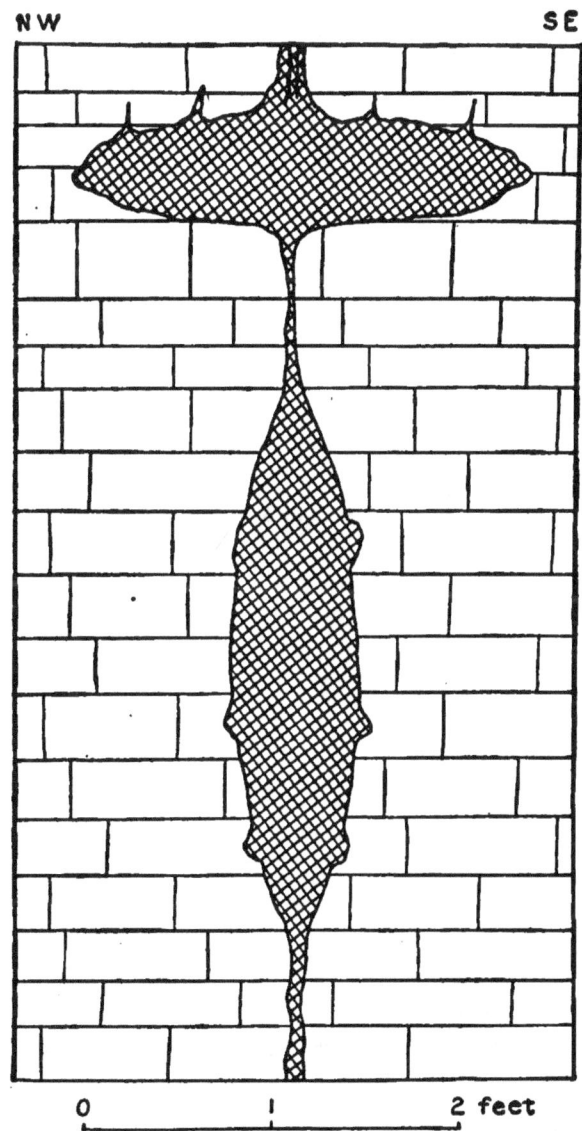

Fig. 48.—Replacement of limestone by argentiferous galena, Bingham, Utah. *After Boutwell.*

being lean or barren. In such cases a study of the stratigraphy is imperative.

Replacement deposits commonly occur in the vicinity of, although not adjacent to, intrusive rocks, and are in some cases closely related to contact deposits; in many instances, however, replacement deposits occur without visible association with intrusives, the mineralizing solutions having gained access to the replaced beds or rocks through fissures.   Replacement deposits unless occurring along a prominent fissure, are likely to be as irregular in distribution through the mineralized horizon as they are individually in form.   Replacement deposits are frequently

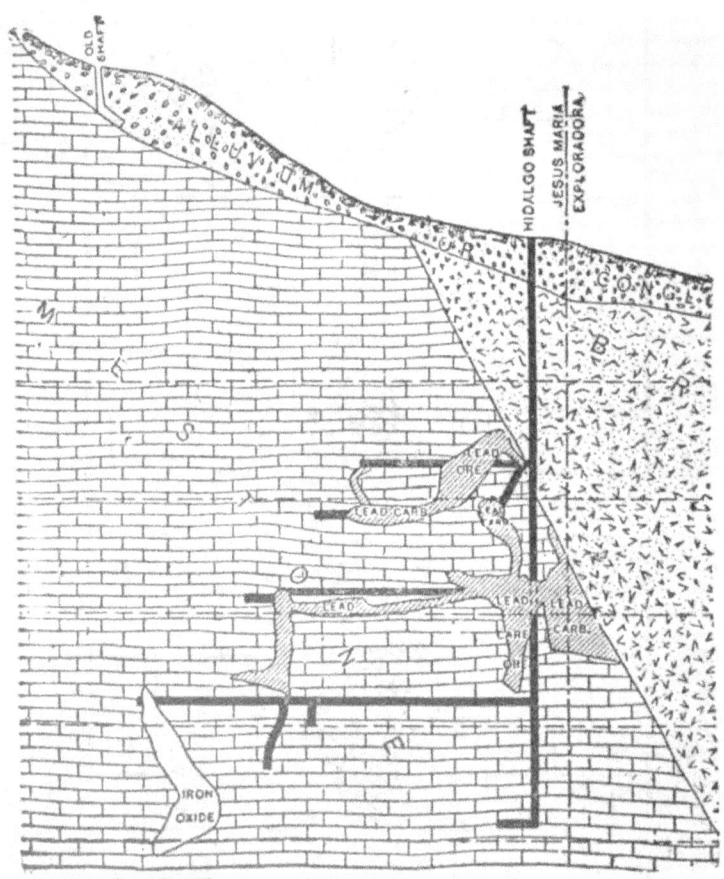

Fig. 49.—Section through the ore-bodies at Sierra Mojada, Coahuila, Mexico, showing irregular replacement deposits in limestone. *After Malcolmson.*

connected with each other or with the main circulation channel, by feeders, or veinlets, which often are quite barren of minerals and with difficulty traceable.   Such obscure feeders are often the only guides in exploration for new deposits.

The degree of shattering, brecciation or straining is, next to the replaceability of the rock, the most important factor in determining the position of replacement deposits; the greater the area of the surfaces exposed in relation to the mass of rock, the more rapidly and thoroughly does replacement take place, and the existence of these deposits may occasionally be predicted through tracing the shattered or strained zones. The passage of solutions through rocks is seen under the microscope to take place even where the existence of cracks or minute fissures is not discernable, and the invading mineral may form grains that are apparently completely surrounded by fresh, solid rock; a zone, therefore, within which the rock has been subjected

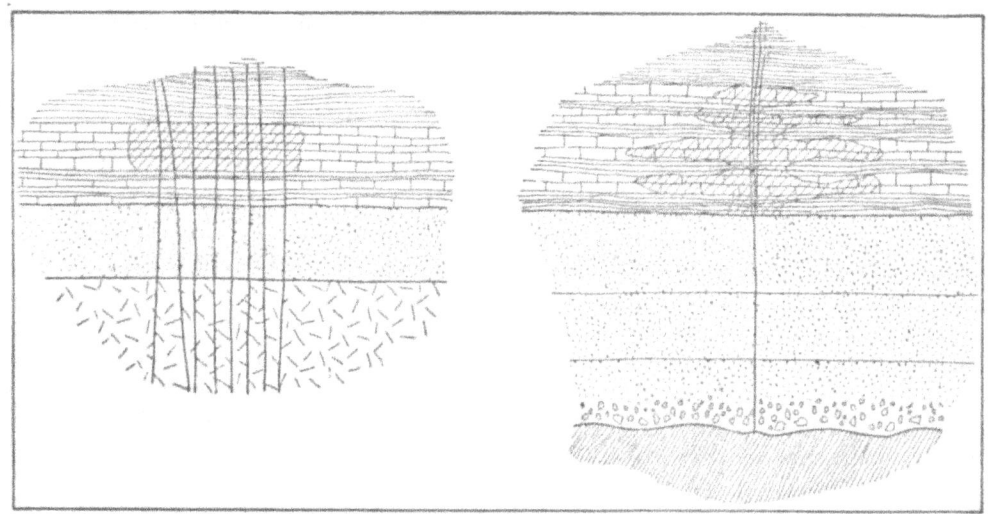

Fig. 50.—Sections showing mineralizing fissures and replacement of limestone strata by siliceous gold ore, Black Hills, South Dakota. *After Irving.*

to slight straining of the particles only, without rupture, may afford access to the replacing solutions. Shattering is a less evident factor in the formation of replacement deposits in limestone than in other rocks, probably because of its ready solubility.

Replacement deposits are rarely of great vertical extent, and very often the rock above them affords no indication of their existence. Outcrops of these deposits, therefore, are accidental, and relatively scarce, and the existence of one such body having been established, blind exploration for new deposits in the same horizon is more often justified than is the case with deposits of other types.

The boundaries of replacement deposits are usually poorly

defined, the ore merging gradually into the enclosing rock, where the boundary between ore and waste becomes a question of assay only.   Occasionally, one mineral has penetrated the rock more easily than the others, and so preponderates in the outer parts of the deposits; in exploration, therefore, upon meeting such a lean, or perhaps barren, mineralization, it may indicate the existence of payable ore beyond.

Replacement ores frequently reproduce the structure of the replaced rock, certain bedding planes being followed in preference to others, and where this condition obtains the bedding planes are usually parallel to the greatest dimension of the ore-body.

The principal alterations of limestones as an accompaniment of mineralization are silicification, marmorization, and to a less extent, the development of a disseminated pyrite mineralization, any of which may afford clews to the existence of ore-bodies.   The primary alterations that accompany mineralization will be discussed in a later chapter.

IN THE COEUR D'ALENES, IDAHO,[1] large replacement deposits have formed along shattered zones in quartzite.   In the Bunker Hill mine the ore consists of galena and siderite, with small quantities of quartz, zincblende and pyrite, the gangue being the enclosing quartzite.   It appears that siderite first replaced the quartzite, and that later argentiferous galena partially replaced the siderite, though the direct replacement of quartzite by galena is occasionally noted.   The best ore consists of rather fine-grained masses of galena with subordinate siderite, which grades into ore in which the siderite exceeds the galena, and this into barren quartzite.   The ore is principally a replacement of the Revett quartzite, but the replacement is closely connected with fissuring, and some of the galena was deposited in open spaces. In some of the important stopes, quartz and pyrite are usually most conspicuous in the transition zone from ore to country rock. The zone of fissured quartzite in which the ore-bodies occur has a maximum width of 300 ft. measured perpendicularly to the Bunker Hill fissure.   Within this zone, here in contact with the

[1] F. L. Ransome, *P. P.* 62, U. S. G. S., p. 162.

foot-wall, there separated from it by barren quartzite, are numerous irregular ore-bodies, usually without definite walls or boundaries. Individual ore-shoots reach 500 ft. in length, 100 ft. or more in width, and 300 to 400 ft. in depth. The whole fissured zone may, in a broad sense, be regarded as a single great lode, within which the partly overlapping and partly connected ore bodies are not uniformly distributed, but are grouped in at least four fairly distinct shoots. That no ore should have been deposited beneath the persistent seam of dark gouge characteristic of this fissure is remarkable, as the quartzites of the foot-wall, which have been well explored, are identical in character with those of the hanging wall, and are in places extensively fissured and broken, though usually to a less degree than in the hanging wall.

In the Highland Boy Mine, Bingham, Utah,[1] important replacement deposits of copper ore occur in limestone. The limestone is commonly a coarsely crystalline marble, in general more cherty toward the base where it rests upon quartzite, more massive and crystalline above, and locally siliceous. The ore bodies lie within the main body of the limestone well above the underlying quartzite along a zone of fissuring and mineralization. Localization of ore has resulted in the formation of three well-defined lenses or shoots, the largest of which reaches a width of 400 ft. and a thickness of 100 ft.; the shoots are approximately conformable to the bedding. In the few instances where cross-cutting has exposed the quartzite, the ore does not make down to it. The walls of the ore-bodies are commonly slip planes or beds of siliceous or crystalline limestone. In some instances the upper parts of the ore-bodies become progressively leaner until they pass into the barren limestone that forms the hanging-wall; laterally, the ore bodies pinch ito thin, irregular seams. The ore consists of pyrite, chalcopyrite with some bornite and chalcocite (secondary?) associated with small quantities of galena, specularite, marcasite, enargite and zincblende; galena is practically restricted to fracture zones. The ore carries small but

[1] J. M. Boutwell, *P. P.* 38, U. S. G. S., p. 267.

commercially valuable quantities of gold and silver. In the exploration for ore-bodies in this district (*Ibid.*, p. 154) on approaching a shoot of copper sulphide ore that lies within barren marble, lean ore may be observed in certain beds. This gradually becomes larger in proportion to the barren rock, and higher in grade, until the entire bed or beds are ore. In the extreme outer parts of these shoots bands of country rock alternate with bands of lean ore, and within the shoot the original bedded character is sometimes preserved by bands of barren siliceous material, and in some cases the massive ore itself preserves the bedded structure.

IN SHASTA COUNTY, CALIFORNIA,[1] large masses of pyritic copper ore have formed as replacement deposits in alaskite porphyry and also occasionally extend into shale; the largest of these deposits, the Iron Mountain, probably originally contained 20,000,000 tons of pyritic ore. The ore-bodies are, in general, roughly tabular, and although irregular in form, may best be referred to as lenses. The ores, which consist of pyrite, chalcopyrite, and zincblende with subordinate galena associated with quartz, calcite and barite, chlorite and sericite, commonly merge gradually into the surrounding rock. The ore-bodies occur in zones of highly shattered and comminuted rock, and this condition is apparently the determining factor in their localization. That the chemical composition of the enclosing rock was not the controlling factor in deposition is proved by the replacement in different deposits of alaskite, shale, and of a basic dike.

IN THE BLACK HILLS, SOUTH DAKOTA,[2] in the Bald Mountain District, large replacement deposits of siliceous gold ores have formed in limestone. Long, narrow, restricted fissures exhibiting, in general, a common trend, pass upward through a series of shales, limestones and quartzites. Where these fissures intersect the limestone, the ore, consisting of pyrite and quartz carrying gold and silver, replaces the rock for considerable distances either side of the mineralizing fissure; where many fissures are grouped

[1] L. C. Graton, *Bull.* 430, U. S. G. S., p. 89.
[2] J. D. Irving, *Economic Geology*, Vol. III, p. 149.

together the mineralization from them has coalesced to form flat masses of great lateral extent. The mineralization along the fissures themselves is commonly slight, and is often absent, and the fissures are so small as to be detected with difficulty in many instances.

At Santa Eulalia, Chihuahua, Mexico,[1] the ore-deposits form great masses of irregular form in a limestone dome, following, and in part limited by, the stratification planes. The ore-bodies are in some instances connected by fissures, by films of red clay, or by limestone checked and netted with minute fractures filled with iron oxide. The ores are for the greater part oxidized, and consist of more or less impure cerussite, sometimes containing cores of residual galena. The replacement of the limestone is indicated by lines of chert through the ore that correspond to similar lines in the unaltered limestone walls, and by the presence of silicified fossils in the ore. The district is one of the most important in Mexico.

Disseminated Mineralizations.—An important class of ore-deposits is that in which the valuable minerals occur as minute particles, or narrow seamlets, or stringers, throughout a large mass of enclosing country rock. The number of such mineralizations whose primary ore is of payable grade is probably small, but these deposits, especially those that contain copper, are of the greatest importance where enriched by secondary processes. Disseminated mineralizations are probably due in great part to metasomatic replacement, but the occurrence of their minerals as sparse disseminations, or impregnations, which are structural terms, is sufficient to warrant their description as a separate type.

Disseminated mineralizations are most frequent in schists and in intrusives; these rocks appear to favor disseminated mineralizations in much the same way that limestone appears to induce segregation and localization of introduced minerals. A characteristic feature of disseminated deposits is the occurrence of the most thorough mineralization in the areas most fissured

[1] W. H. Weed, "Nature of Ore Deposits," Beck-Weed, p. 573.

or shattered. In many cases the mineralization accompanies reticulated quartz veinlets through the shattered rock, and often, especially where the dissemination occurs in the mineralizing intrusive, the introduced minerals are abundant along joint planes as well as fracture planes, and occur in much less quantity in the interior of the masses bounded by such surfaces. Disseminated mineralizations are commonly closely associated with intrusives, and where important, extend through very large rock masses.

At Bingham Canyon, Utah,[1] a great mass of intrusive monzonite carries throughout an irregular but persistent mineralization of finely disseminated pyrite and chalcopyrite that carry low values in gold. In the fresh monzonite these minerals occur as minute grains scattered through the rock, and along joint planes, and also embedded in irregular quartz veinlets. A correlation of assays with structure indicates that the values are highest where the fissuring and veining are most pronounced. Through secondary enrichment this deposit has yielded copper-deposits of the first rank.

At Clifton, Arizona,[2] intrusive monzonite-porphyry carries a disseminated mineralization of pyrite and chalcopyrite, with subordinate zincblende and molybdenite, associated with quartz and a sericitization of the containing rock. This mineralization is most intense in and along certain veins through the monzonite-porphyry, but extends as impregnations through the rock, along joint planes, and associated with quartz veinlets, for long distances either side of the mineralizing veins. The containing rock is here, also, the mineralizing intrusive. Large areas of the intrusive carry sulphides, but the mineralization is apparently most intense in the vicinity of the centers of intrusion at Morenci and near Metcalf. The primary ore is unpayable, but through secondary enrichment it has yielded important deposits.

In the Burro Mountains, New Mexico, an intrusion of monzonitic porphyry through granite has been accompanied by a dis-

[1] J. M. Boutwell, *P. P.* 38, U. S. G. S., p. 259.
[2] Waldemar Lindgren, *P. P.* 43, U. S. G. S., p. 202, 222.

seminated mineralization consisting of minute grains and seamlets of pyrite and chalcopyrite associated with quartz. The mineralization is most intense along certain fracture zones, and appears to be proportional to the amount of shattering of the enclosing rock. The primary mineralization is unpayable, but has at two localities yielded important deposits through secondary enrichment.

AT THE HOPEFUL MINE, NOGAL DISTRICT, NEW MEXICO, a much altered rock, probably sericitized and later kaolinized, carries a disseminated pyritic mineralization containing gold but no copper. The pyrite occurs as grains through the rock, which L. C. Graton states to be probably a monzonite, and is most abundant along joint planes and certain ill-defined fissures. The gold, which is stated to vary between $1.00 and $3.50 per ton, is said to be uniformly distributed, and to be approximately equal in the oxidized and in the sulphide ore.

Conglomerate Beds.—While not a numerically important type, mineralized beds of conglomerate form the ores of two of the most important mining districts in the world. The origin of these deposits is not clear, and while to a certain extent similar, they posses features that render difficult their classification with the more common and better understood deposits. In both the Michigan copper deposits and the gold deposits of the Rand the ore consists of native metals in the cementing material of conglomerate beds, and in both districts the mineralizations are remarkably persistent over great areas, and to great depths.

IN THE WITWATERSRAND, SOUTH AFRICA,[1] beds or "reefs" of conglomerate are intercalated with quartzitic sandstones and, more rarely, slates; there are eight groups of these reefs, certain of which have been exploited over a length of 48 miles. In thickness the reefs vary between a maximum of several meters to a complete wedging out. The thinner reefs are commonly the richer. The intervening strata are practically barren, although exceptions to this rule are known. The reefs are composed of pebbles, commonly ranging from a hazel-nut to a hen's egg in

[1] Beck-Weed, "Nature of Ore Deposits," p. 512.

size, of quartz and quartzite, more rarely of siliceous schist, and occasional rounded pyrite granules; the pebbles are often deformed, being flattened or splintered, and themselves rarely carry any mineralization. The cementing material is composed of small quartz granules and pyrite with associated particles of gold. The pyrite occurs in rather irregular distribution, and frequently forms crusts around the pebbles of the conglomerate, and occasionally is concentrated in delicate films parallel to the stratification. The mineralization, while exhibiting local irregularity, is relatively uniform and persistent over long distances both in strike and in depth.

ON KEWEENAW POINT, MICHIGAN,[1] beds of copper bearing conglomerate occur interstratified with sandstones and sheets of diabase, both compact and amygdaloidal, and with melaphyre; certain beds of strongly altered diabase scoriaceous in character are known as ash beds. In the conglomerates, the copper has replaced the finer particles so as to appear as a cement; the boulders themselves, or particular minerals in them, are often permeated with copper, which occasionally occurs in large masses. The copper is associated with chlorite, epidote, and abundant zeolites. The associated amygdaloidal rocks carry copper in their small cavities, and in certain shattered areas it occurs irregularly, occasionally in fragments of large size. The rich parts of the beds occur as shoots, in some instances several thousands of feet in length; these shoots continue somewhat diagonally down the dip of the beds to great depths without essential diminution or change in mineralization. The distribution of the copper in the amygdaloidal sheets is much the same as in the conglomerate beds.

Bedded Ore-Deposits.—Where mineralization has proceeded contemporaneously with the deposition of the enclosing bed the resulting deposit is known as a seam, or bedded deposit; in this class are also included those replacements of similar occurrence the origin of whose mineralization is not apparent. The criteria for distinguishing between bedded deposits and intercalated veins

---

[1] J. P. Kemp, "Ore Deposits," p. 204.

are given in a preceding paragraph. Beck makes a further distinction[1] between interbedded deposits, which are overlain by other strata, and superficial deposits where there are no overlying beds, as, for example, beds of bog iron ore.

Bedded deposits are occasionally recognizable by contained fossils, which may have become mineralized. These deposits, of which a majority contain iron or manganese, are commonly of large horizontal dimensions as compared with thickness, and where the strata are folded, they follow all the sinuosities of the containing bed; in occurrence they are comparable with coal seams. A study of the stratigraphy of the bed containing the ore and of the associated strata is imperative in the investigation of such deposits.

Certain types of bedded deposits terminate through wedging out around their peripheries, others toward their edges become gradually poorer through the occurrence of barren partings, which increase in proportion to the ore until the mass becomes unpayable. Bedded deposits whose mineralization was contemporaneous with the deposition of the containing stratum are commonly persistent over large areas.

Bedded deposits whose mineralization is later than the deposition of the containing bed are less likely to be persistent over large areas, and their contained mineralization is likely to have been controlled by some constituent of the bed, such as carbonaceous material, and to fail over such parts of the bed as did not contain such preciptants. Where two or more bedded deposits occur in the same series they are persistently parallel through all the sinuosities of the associated strata.

THE CLINTON ORE MEASURES OF THE UNITED STATES,[2] are geographically persistent in extent, and wherever they outcrop they almost invariably contain one or more beds of red hematite intercalated with sandstones and shales. In occurrence the ore varies somewhat, at times being the replacement of fossils, again as small oölitic concretions, and in places constituting a highly

[1] Beck-Weed, "Nature of Ore Deposits," p. 49.
[2] J. F. Kemp, "Ore Deposits," p. 115.

ferruginous limestone. In some cases it can be shown that
these beds result from the weathering near the surface of beds
of ferruginous limestone, and are thus secondary or residual in
character. They constitute deposits of great economic importance in several states.

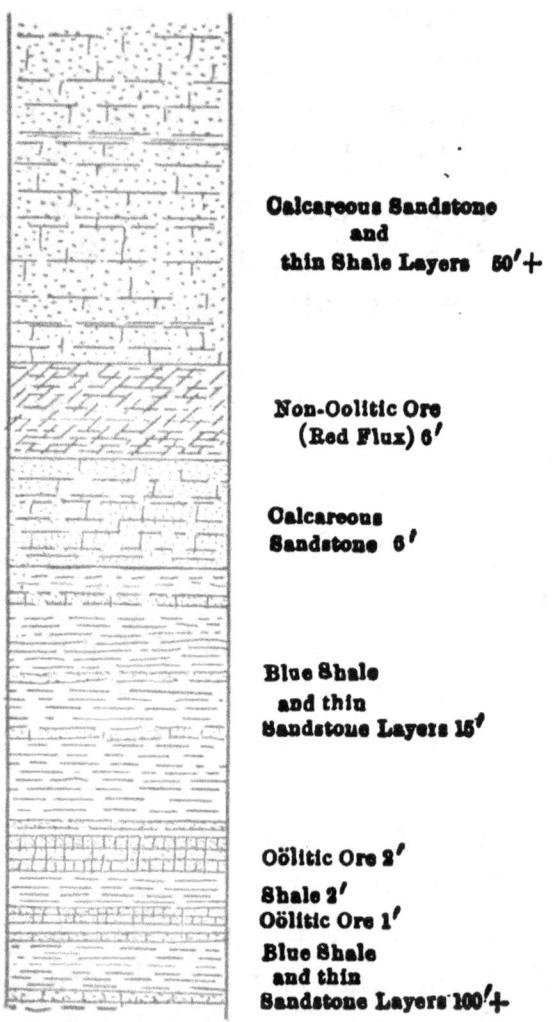

Calcareous Sandstone
and
thin Shale Layers 50′+

Non-Oolitic Ore
(Red Flux) 6′

Calcareous
Sandstone 6′

Blue Shale
and thin
Sandstone Layers 15′

Oölitic Ore 2′
Shale 2′
Oölitic Ore 1′
Blue Shale
and thin
Sandstone Layers 100′+

Fig. 51.—Section of the Clinton ore measures, Clinton, New York, showing
the bedded iron ores. *After C. H. Smith.*

Near Mansfeld, Germany,[1] a stratified copper deposit extends
over an area 120 miles long and 60 to 90 miles wide; the average
thickness of the bed varies between 18 and 23 inches; the copper
bearing member is a blackish, bituminous shale which lies un-
conformably upon red sandstones and conglomerates, and is
overlain by limestones and dolomites. The copper occurs in the

[1] Beck-Weed, "Nature of Ore Deposits," p. 488.

form of chalcopyrite, bornite and chalcocite, associated with pyrite, galena, zincblende and other sulphides, and contains a little silver. Although the entire bed is copper-bearing, the payable ore is commonly confined to a layer from 3 to 6 inches in thickness, and appears to be associated with the bituminous material, the bed becoming leaner in its lower portion where the carbonaceous material becomes less in quantity; replacements of fossils, especially those of fishes, are frequent. The average value of the ore appears to be from 2 to 3 per cent. copper, with some silver.

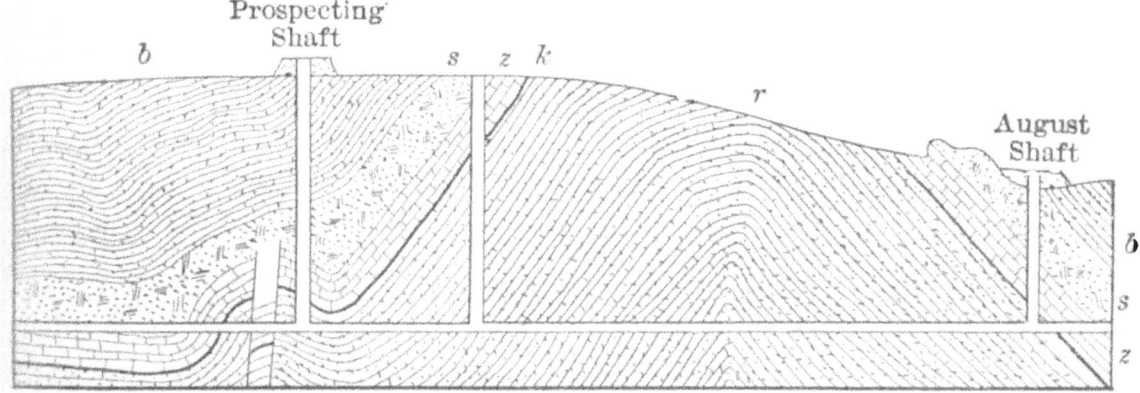

FIG. 52.—Section of a partly eroded anticline near Mansfeld, Germany. showing the Mansfeld copper shale as a black line. *After Schrader.*

THE RED SANDSTONE BEDS OF THE SOUTHWESTERN UNITED STATES.—These beds at many places carry copper as chalcocite with a little pyrite and chalcopyrite replacing organic matter, such as tree trunks, leaves, bark, and associated finer particles. Occasionally, the kaolin or calcite of the cementing material between the grains of the sandstone is replaced. These deposits appear to be unconnected with fissures and, except at Tularosa, N. M., are not in association with intrusives. The chalcocite, which commonly carries a little silver, is evidently of later origin than the containing beds, but the source of mineralization is not apparent. These deposits, while frequently confined to a single bed, are commonly present in several beds of the sandstone series. The mineralization is extremely irregular, and these deposits have not yielded satisfactory results on exploration.

8

CHAPTER VI

PRIMARY ORE-SHOOTS

A preceding paragraph treats of the irregular manner in which ore-deposits occur and the complex factors that control their distribution; the occurrence of metals in ore-shoots in individual deposits is equally irregular, and the factors that control such segregation are equally complex. While the dimensions of ore-shoots in individual deposits may not be foretold with any degree of accuracy, the occurrence of ore in shoots is not accidental, but is controlled by laws some of which are understood. Any one familiar with permutations and combinations who stops to consider the variety of factors that control ore deposition, and their varying relative importance, will admit that the science of economic geology must advance greatly before the occurrence of ore-shoots may be predicted accurately in advance of exploration.

The Factors that Determine Primary Ore-Shoots.—The segregation of minerals in any ore-shoot must be referred to some condition, or combination of conditions, local to the ore-shoot as compared with the remainder of the deposit of low-grade or barren minerals. Among the localizing factors that have been recognized are, in filled fissures, the amount and distribution of open spaces available for ore deposition, and in replacement deposits, the degree of brecciation giving access to the replacing solutions; the intersection of veins with other veins, dikes, sheeted zones, or porous strata; the impounding of solutions by impervious strata; and the differing precipitative influences of enclosing rocks. A large proportion of important and well-defined ore-shoots, especially in deposits of deep-seated origin, are not assignable to any of these factors.

The Relative Value of General and Local Data.—While a

114

knowledge of the character of ore-shoots in general is important, a knowledge of the behavior of the known ore-shoots in the property or in the district under investigation is even more significant. Accurate and complete assay and geological maps of the areas already mined are essential to an intelligent investigation of ore-shoots, and it is strange that such maps are so rarely to be found at even important mines.

The distribution of ore in shoots is the greatest factor in the risk of mining, and the factor of most vital importance to the mine owner and mining engineer. Not infrequently, a practical man from long experience in a particular deposit can predict its behavior in advance of exploration with a valuable percentage of accuracy; this reliability of prediction would in many cases be greatly increased by a technical study of complete assay and geological maps, upon which relations become definite that otherwise would not suggest themselves.

Ore-shoots vary with commercial as well as with natural conditions, and material left in place as waste may in later years constitute a valuable ore. Furthermore, while the difference between assays of 20 cents and $1.00 is not of immediate commercial importance as regards the marketability of an ore, it may be of vital importance as a guide to a new ore-shoot. It is natural to extract ore and to be satisfied as long as the ore lasts, but all ore-shoots come to an end, and then, when a search is commenced for further reserves, the data of most value in conducting that search will have been lost unless carefully recorded on a map, and the exploration must be conducted by a process of elimination which may or may not yield results.

Primary and Secondary Ore-Shoots.—In the study of any deposit the first question to decide is whether the ore-shoots are due to primary segregation of values or to secondary enrichment by surface agencies.

It is much easier to recognize a secondary ore with certainty than to determine an ore to be certainly primary, for the reason that the criterion of a primary ore is the known lack of secondary additions. With a majority of ores microscopic study in thin section will reveal primary character, but there are many instances where the assumed primary nature of an ore is open to question, especially with ores of gold or silver.

Even below the zone of recognized secondary enrichment, deep exploration usually shows a falling off in value in the primary ore in depth, and the actual influence of surface agencies may extend to greater depths than is generally supposed.   It often happens that commercially valuable ore ceases at the depth reached by secondary agencies, and the data is likely to be scanty in

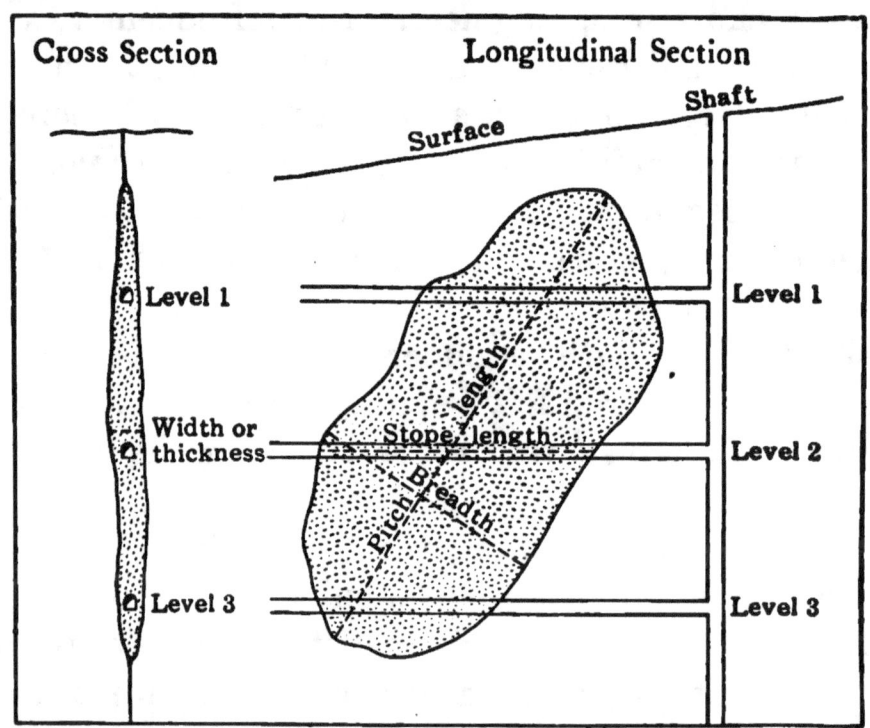

Fig. 53.—Diagram illustrating the terms used to describe the dimensions of ore-shoots.   *After Lindgren and Ransome.*

regard to the primary distribution of values where not obscured by surface agencies.   Secondary ore-shoots will be considered in a succeeding chapter.

Terms Used to Describe the Dimensions of Ore-Shoots.—The most convenient terms for use in describing ore-shoots are those suggested by Messrs. Lindgren and Ransom.[1] The longest dimension of a shoot is called the "pitch length"; the horizontal length the "stope length"; the width at right angles to the

[1] *P. P.* 54, U. S. G. S., p. 205.

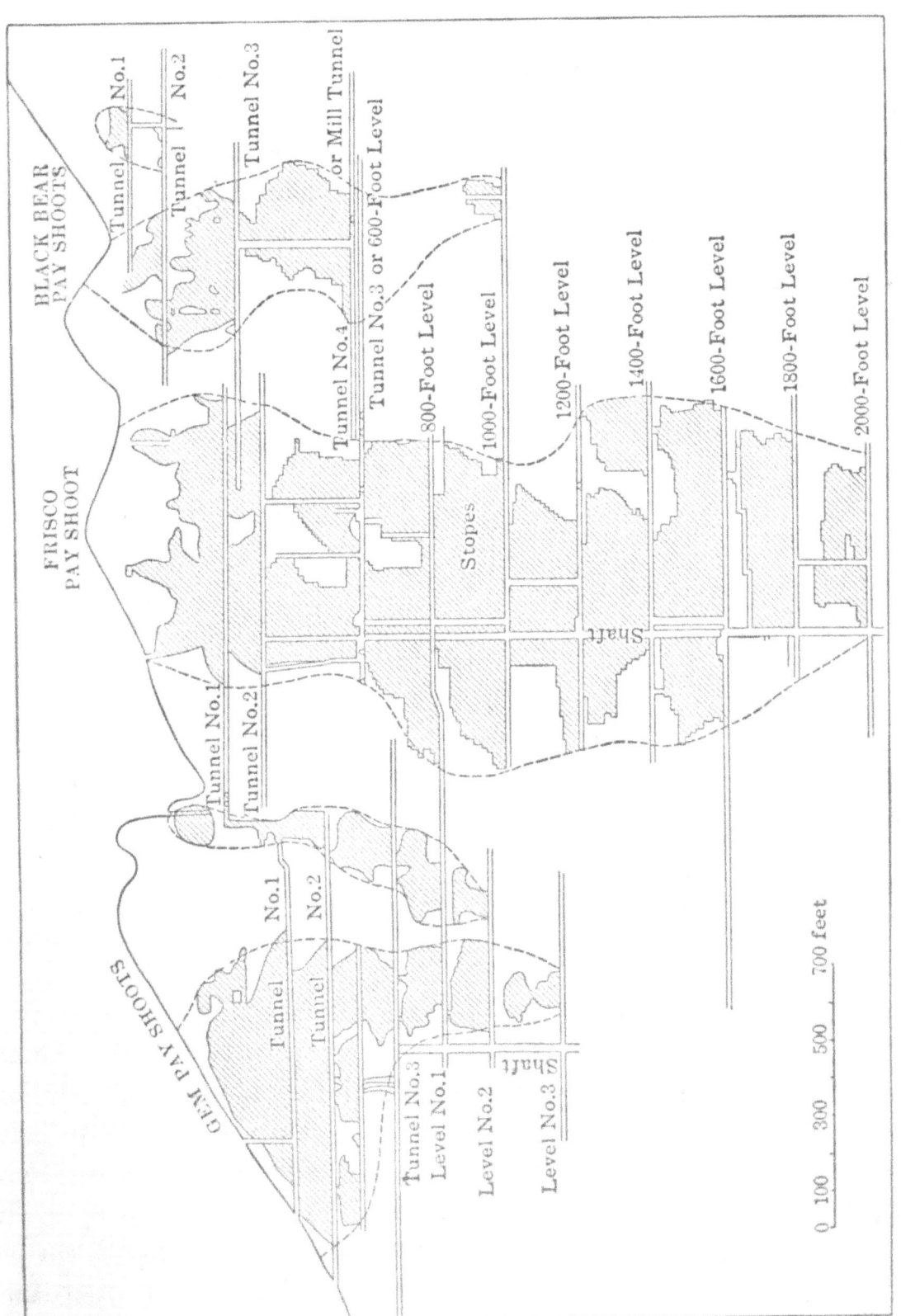

FIG. 54.—Primary ore-shoots in the Helena-Frisco mine, Coeur d'Alenes, Idaho. *After Ransome.*

pitch length is termed the "breadth"; thickness is measured at right angles to the plane of the pitch length and breadth.

The Shapes of Ore-Shoots.—Probably no mineral lode approaches the regularity of a seam of coal, although as compared with the majority of mineral veins some are notably regular. The Smuggler vein at Telluride, Colorado, and the older, East-West,[1]

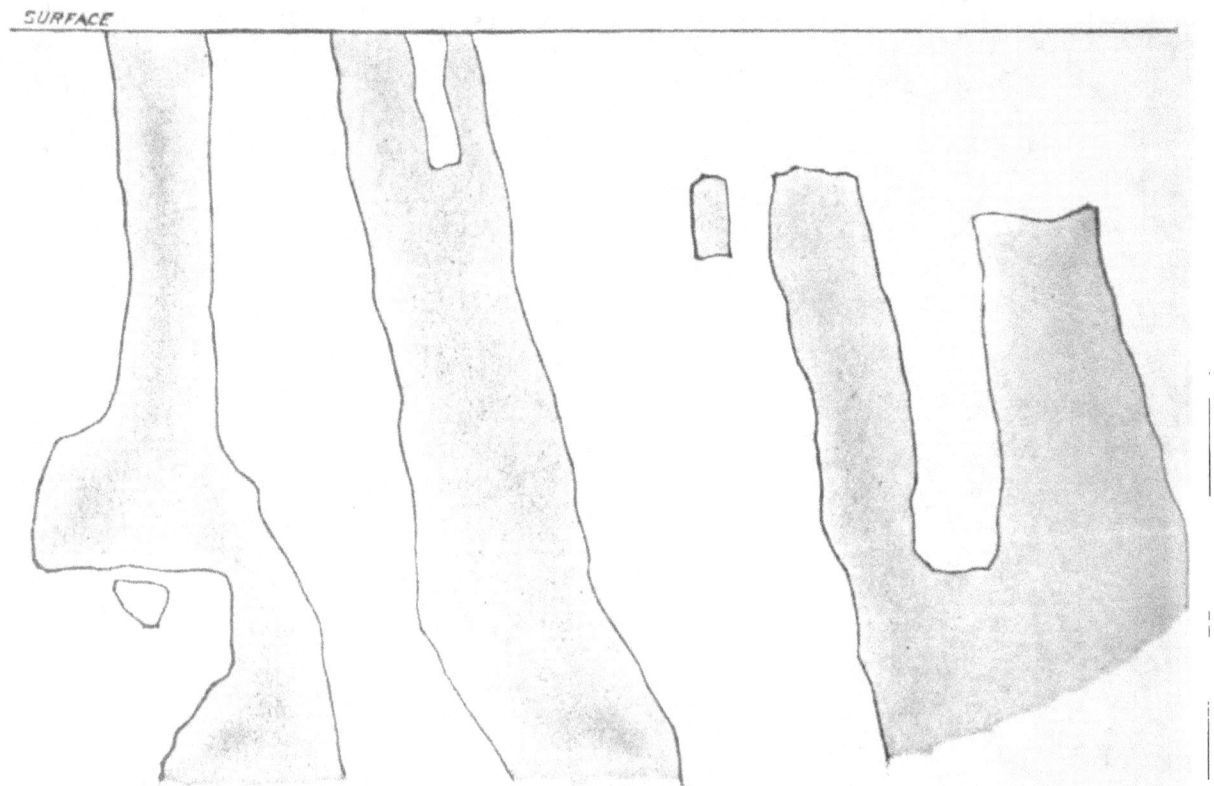

SURFACE

FIG. 55.—Primary ore-shoots, Grass Valley, California. *After Lindgren.*

veins at Butte, Montana, are remarkably regular and uniform in mineralogical character; the more prominent of these veins have been stoped for thousands of feet along their strike, showing little, if any, disposition to develop shoots; the gold reefs of the Rand, South Africa, are markedly regular over long distances as compared with most lodes. These examples, while of marked general regularity, show considerable variation in value at different points. The other extreme, that of irregularity, is more common, and instances are numerous where the

[1] R. H. Sales, *Economic Geology*, Vol. III, p. 327.

valuable metals occur in small high-grade masses so irregularly distributed through the gangue that they may not be developed ahead of extraction, but must be discovered by chance and mined out as found. Between these two extremes may be classed a majority of ore-deposits, where the payable ore occurs in fairly definite shoots, the size and shape of which depend upon the market prices of metals and the cost of mining as well as on geological conditions.

The broadest generalization that can be made in regard to the shape of primary ore-shoots is that their vertical dimension is likely to exceed, even considerably exceed, their horizontal dimensions: in secondary ore-shoots the reverse is the rule. There are many instances where primary ore-shoots are as well defined in vertical as in horizontal extent, as, for example, are many of ore-shoots at Cripple Creek, Colorado;[1] here the ratio between pitch length and breadth varies from 1 1/2 to 1 to 5 to 1 in the shoots that have not been truncated by erosion.

Some generalizations have been attempted in regard to the relative size of ore-shoots at the surface and in depth: this applies with some force to secondary ore-shoots and to those primary ore-shoots whose upper parts have been leached. It is evident, however, that most primary ore-shoots were formed at very considerable depths and that the present surface must be considered accidental and therefore not a factor in determining the primary distribution of metals. Where primary ore-shoots are definitely limited in vertical extent they are likely to be of roughly lenticular form, and other shoots are likely to appear in the projection of their pitch lengths in depth. In some districts primary ore-shoots are persistent to great depths, although of well-defined breadth.

No valuable generalizations can be formulated, but there is often an approach to regularity in the relationship among the ore-shoots, of individual mines or districts, as regards pitch, continuity, relative pitch-length to breadth, and in the regularity of sequence of different shoots. The effect of structural con-

[1] Waldemar Lindgren and F. L. Ransome, *P. P.* 54 U. S. G. S., p. 205.

ditions upon ore-shoots and the chemical effect of the wall rocks in inducing precipitation will be taken up in a succeeding paragraph.   Where these factors do not appear to have had effect in the distribution of metals, the causes of their segregation into shoots are not understood, unless they are referred to mass action, where a precipitation once started, by whatever cause, becomes a continuous process until the final result is an ore-shoot.

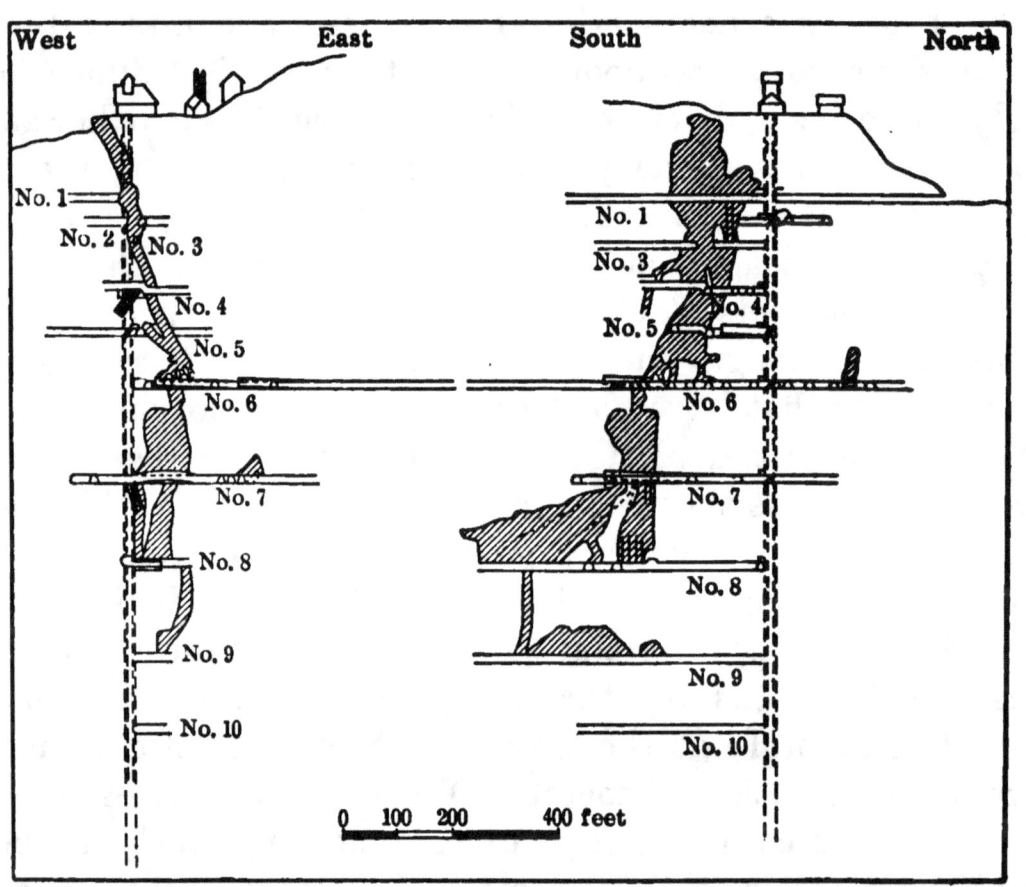

Fig. 56.—Sections of the Silver Bell ore-shoot, Silverton, Colorado.   *After Ransome.*

A certain type of ore-shoot gradually fades out around its peripheries, the values becoming progressively less from the center toward the edges, where the final boundary of the shoot is determined by the limiting cost of mining and treatment This fading out may be accompanied by a visible change in the mineralogy of the shoot, or the same gangue and acessory minerals may persist beyond the payable values; the vein may pinch out with the diminishing values, or may continue as strongly beyond them. Another type of shoot presents sharp outlines, the grade of ore being maintained up to the boundaries of the shoot.

In many wide, compound lodes or shear zones the ore-bodies occur with great individual irregularity, either connected with each other by stringers or apparently unconnected, and lying either along the foot- or the hanging-wall; taken collectively, however, they are likely to show a general alignment, or arrangement in shoots, and the lode in the direction of their collective pitch lengths may be considered promising territory.

Ore-shoots not infrequently follow an overlapping arrangement where each shoot overlaps the one succeeding on one side and the one following on the other. Where there are several parallel fissures an ore-shoot on one vein is likely to be succeeded by a slightly overlapping shoot in the adjoining vein.

Lenticular Ore-Shoots.—Lenticular masses of ore in schistose rocks constitute another class of primary ore-shoots; here the original ore-body, of whatever origin, has been transformed by regional metamorphism into lenticular masses oriented with their plane of greatest dimension roughly parallel to the schistosity. These lenticular masses are likely to follow one another in overlapping sequence, but often occur singly.

The lenticular form is often well-defined, the central part being thickest, and the ore gradually diminishing in thickness toward the edges, where the lens dies out in the shist.

During the formation of these lenses recrystallization appears to destroy the lines of schistosity through the mass of the ore, but the schistose structure is commonly present near the boundaries, where the ore is often mingled with parallel bands of the minerals of the enclosing schist. In pyritic lenses carrying copper, a part of the chalcopyrite is likely to be segregated and to occupy veinlets through the pyrite.

In many cases lenticular ore-bodies appear to have controlled the schistosity of the enclosing rock, and in the exploration for further lenses a plotting of the schistosity is often helpful. Mr.

W. H. Emmons[1] states that where such deposits are separated to form overlapping lenses, there is some evidence to show that they separate where the ore is most siliceous, for a mixture of sulphides and quartz seems to be less capable of resisting stresses than the more massive pyrite. In such cases a thin fissure with the schistosity parallel on either side will probably be found to connect the broken ends.

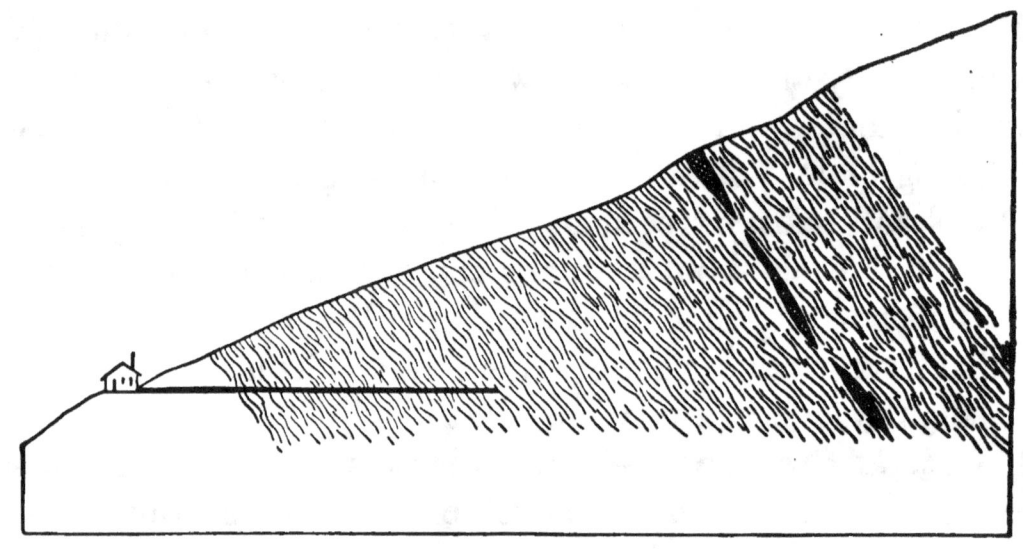

FIG. 57.—Section showing the lenticular masses that make up the Continental vein, Encampment, Wyoming. *After Spencer.*

The Behavior of Primary Ore-Shoots in Depth.—A majority of the ore-deposits of our Western States are believed to be the result of ascending mineral-bearing solutions which deposited their burdens where diminishing temperature and pressure permitted precipitation upon nearing the surface.

Assuming a uniform fissure through a homogeneous rock, it would seem, therefore, that the changes to be expected in depth would be those due to changes in pressure and temperature, the most easily precipitated mineral, or the mineral in greatest excess, extending the deepest, with the other minerals following in regular sequence until near the surface the most soluble mineral would be found. This is partly borne out by recorded facts, but no absolute succession of minerals may be distinguished,

[1] *Economic Geology,* Vol. VI, p. 781.

for the reason, possibly, that the solubilities of the different elements and compounds vary greatly according to the presence of other compounds in solution, and also by reason of the interference of structural conditions and the chemical effects of enclosing wall rocks. The usual order is, nearest the surface, compounds of mercury, then galena, blende; chalcopyrite or cupriferous pyrite, with pyrite the lowest. These zones all grade one into the other, and in many cases but little change is noted in the relative abundance of these minerals over considerable vertical distances. Gold appears to be precipitated at all depths, free gold and auriferous pyrite extending the deepest, and the compounds of gold of greatest mobility, such as tellurides, extending the highest. The position of silver appears to be higher up in the series. In the absence of arsenic, antimony, bismuth and tellurium, silver is usually associated with galena, but in the presence of these elements, it is likely to be higher up in the series and combined with these elements in preference to lead. The primary character of certain of the silver minerals is open to question. This sequence is partially corroborated by the typical associations of the metals. Of the six elements, silver is most commonly associated with galena, galena with zincblende, zincblende with chalcopyrite (to a less marked degree), chalcopyrite with pyrite, and pyrite with gold.

The general tendency is for ore-shoots to become smaller, but more regular, with increase in depth.

The Decrease in Value with Depth.—Primary ores commonly show a progressive decrease in value in depth. Mr. Waldemar Lindgren says[1] "This decrease is likely to be rapid near the original apex of the veins, but below this it is in most cases very slow, extending over a vertical range of many thousand feet." As examples, he gives the Grass Valley Mines, California, where the North Star vein produces a gold ore from a vertical depth of 1600 ft., which corresponds to 4100 ft., on the incline, equally as rich as that found at higher levels. The saddle reefs

---

[1] *Economic Geology*, Vol. I, p. 45.

at Bendigo, Australia, contain payable ore at a depth of 4156 ft. In the Coeur D'Alenes, Idaho,[1] the lead-silver ores appear to maintain their tenor to depths of 1800 ft. below their outcrops without sign of diminishing values.

If any generalization may be made, it would seem that silver, lead and zinc are less likely than gold to be persistent over great vertical distances.

Predicting the Depth to which Ore-Shoots may Continue.—The primary character of an ore being established, and the effect of structural and chemical precipitants not being apparent, the chances are good that an ore-shoot will maintain its values in depth. A shoot whose partial development indicates a widening tendency, or fairly uniform breadth through several levels, has, of course, a much better chance of vertical persistency than one that is definitely narrowing as it goes down, which probably belongs to the type of ore-shoot that has definite vertical as well as horizontal limits. If the variation in value along the dip is no more pronounced than along the strike, there is no basis upon which to presuppose the absence of ore in the deeper portions of the vein, where further exploration may disclose other ore-shoots, most likely in the projection of the pitch length of a shoot higher up on the vein. It has been established, however, that in many instances where the values are maintained in depth, the absolute quantity of ore grows less with increasing depth, which may be a condition more apparent than real, on account of the greatly increased expense of deep exploration.

The Depths at which Ore-Deposits Form.—Ore-deposits have been classified on the basis of the depths, or the conditions, under which they were formed. Deposits of igneous, pegmatitic or of contact origin, and of abyssal, moderate and of shallow depths exhibit characteristic mineral associations, which frequently permit the investigator to assign an ore-deposit to one of these classes.

A knowledge of the depth at which an ore was formed gives an

[1] F. L. Ransome, *P. P.* 62, U. S. G. S., p. 130.

idea of the amount of erosion that took place to expose the deposit at the surface, and permits an intelligent corelation to be made with deposits of similar origin elsewhere.

The Structural Features that Influence Ore-Shoots.—With the exception of magmatic segregations and deposits due to contact metamorphism, a prerequisite of mineralization is the existence of a fissure or fissures to give access to the mineralizing solutions; the mineralizing effect of these solutions is probably controlled to a large degree by the character and changes of these circulation channels.

One extreme of fracturing may be considered the solid, unbroken rock, through which solutions work their way very slowly. The other extreme is a crushed or ground-up condition where the fineness of comminution is sufficient to produce a gouge or clay-like mass, which is also relatively impervious to the passage of solutions. Between these two extremes lie the favorable conditions for ore deposition.

Ore is deposited from solutions either in open spaces or by replacement of the fissured rock. In replacement deposits the degree of mineralization is frequently proportional to the degree of brecciation, as replacement proceeds most rapidly where it has the greatest area of rock surfaces on which to work, until the point is reached where the fineness of comminution commences to retard the passage of the solutions. There is, however, a point where brecciation ceases to be an advantage—where it is so extensive that the solutions are dissipated through a large mass, and the resulting mineralization is too scattered to yield a commercially important deposit.

It sometimes happens that two or more systems of fracturing or brecciation have effected the rocks; here the time of the fissuring becomes important. Fissures heal and become closed in time, and so cease to afford circulation channels; the most favorable time of fissuring or brecciation may be considered that just before the introduction of the solutions—perhaps due to the same igneous disturbances to which the solutions owe their origin. Post-mineral fracturing is, of course, of no primary effect, but may be of the greatest importance in working secondary changes.

In the investigation of a mineralized area, therefore, the distribution and extent of the brecciated masses, and the degree and relative time of brecciation should be studied; mineralization is frequently co-extensive with or confined to the brecciated areas. The brecciated structure of an unmineralized rock is sometimes not visible on fresh fractures, but becomes plain on weathering, or upon being wetted.

The degree of brecciation varies greatly in different rocks; massive or rigid rocks yield breccias where soft or plastic rocks yield to strain without brecciation. The continuity of a brecciated zone from one rock into another is, therefore, uncertain.

The degree and character of brecciation is of the greatest importance in the study of the surface exposures where the existence of disseminated copper ores is suspected.

AT BINGHAM, UTAH, Mr. J. M. Boutwell,[1] referring to the disseminated mineralization of the great laccolithic mass, states that: "In general, underground observations tend to show that chalcopyrite and pyrite occur in greatest quantity where the country rock has been most broken."

AT ASPEN, COLORADO, referring to the silver lodes, Mr. J. E. Spurr[2] states that: "A microscopic study of limestone in the process of replacement by dolomite, silica, and sulphides shows that the rock was first profoundly strained and crushed in the vicinity of faults, so that many tiny passages were opened to solutions, which finally worked through and through the strained material, replacing it. The amount of straining, which regulated the area of surface offered to the solutions, usually determined whether there would be little or much replacement; where the circulation zone along a vein is wide and is filled largely with finely ground material which is not so pasty as to prohibit free circulation through it, the replacement of earthy materials by metallic minerals as well as their accompanying

[1] *P. P.* 38, U. S. G. S., p. 131.
[2] *P. P.* 63, U. S. G. S., p. 160.

gangue goes on very much more rapidly than elsewhere along the vein, where the rock is hard and is only sliced by fracture planes."

Ore-Shoots due to Available Open Space.—In deposits formed by the filling of open spaces the distribution of these open spaces is usually the controlling factor in ore deposition, and the segregation of metals into ore-shoots bears the closest relation to the vein structure.

The pinching and swelling of veins in fissures of small displacement have been discussed in a preceding chapter; where this condition can be shown to exist, the pinching out of an ore-shoot may be reasonably expected to be followed by at least a widening of the vein, which may or may not contain ore.

A vein that is the result of both replacement and of the filling of open spaces is likely to vary in mineral content according to which of the two processes took place during ore deposition, the open spaces perhaps contributing the ore-shoots while the replacement of the walls where the fissure was tight is represented by intervals of barren or low-grade filling.

In a crustified vein, if it is determined that the valuable mineral was among the first formed, and relatively near the walls, the ore is likely to be persistent as compared with a mineral that was among the last to form and so confined to such spaces as were kept open the longest.

Ore-Shoots due to Intersections.—Ore-shoots are frequently developed where a vein is intersected by another vein or by a fissure, dike, sheeted zone, or branch vein. This association is more common in deposits that have been formed near the surface than in the deposits of deep seated origin.[1]

Intersecting members appear to affect the conditions of circulation, or to introduce chemical or physical changes that accelerate the deposition of ore minerals along such junctions.

An intersection may afford, through increased local crushing, a more open channel for the passage of solutions, and so lead to

---

[1] See *Economic Geology*, Vol. I, p. 43, Waldemar Lindgren.

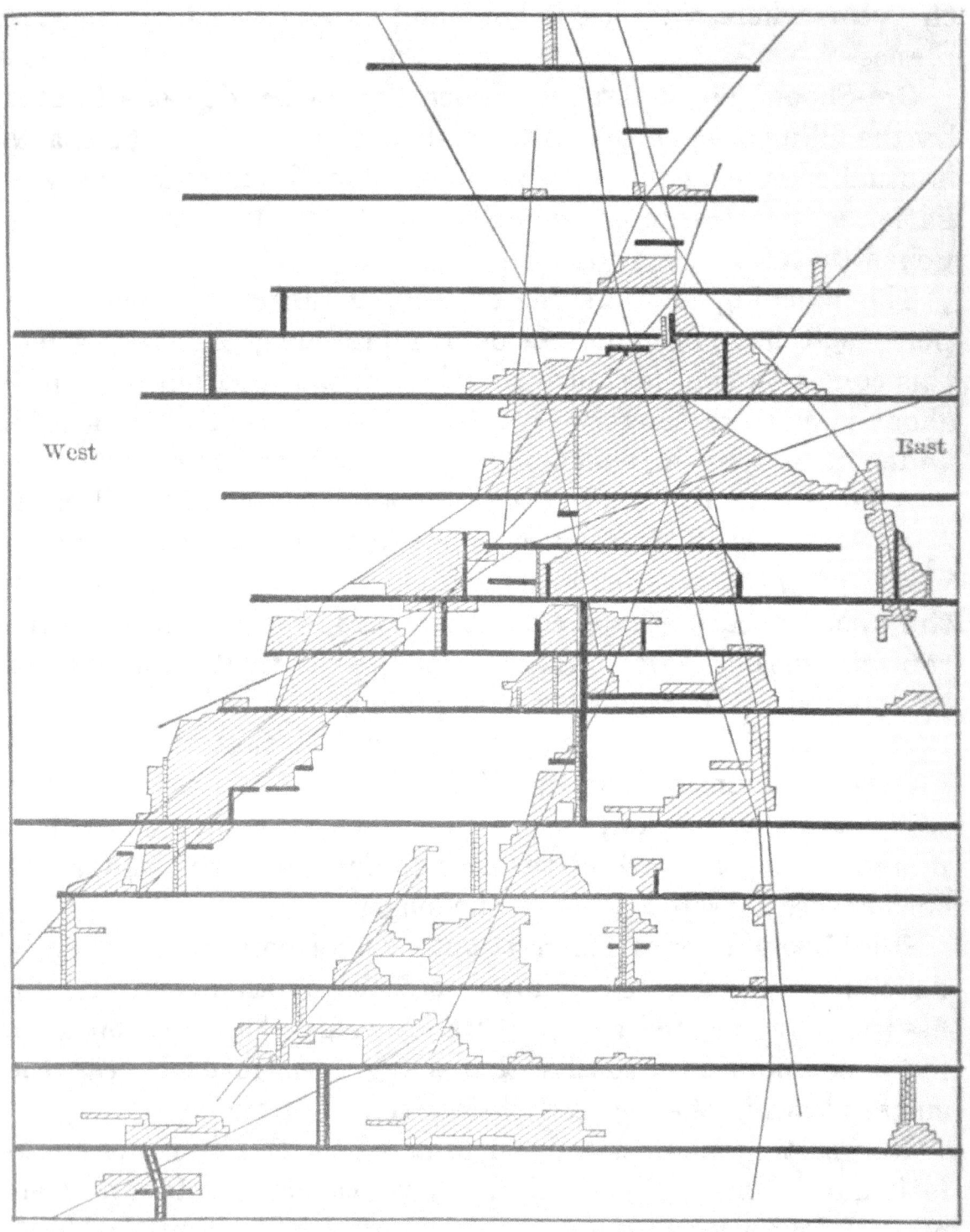

Fig. 58.—Longitudinal section along the Neu-Hoffnung vein, Himmelsfahrt, Germany, showing the ore-shoots along the intersections with several other veins. *After Beck.*

the passage of a greater quantity of the mineralizing solutions through the part of the vein affected by the intersection; moreover, a partial stoppage of circulation appears to induce precipitation, stagnation, or partial stagnation, being favorable to precipitation. In some instances the intersecting veins carried solutions of different metal content, as is evidenced by their different vein fillings, and through a mingling of the two solutions chemical precipitation was brought about that formed ore-shoots: furthermore, a mingling of solutions under different conditions of temperature or pressure, appears likely to disturb the equilibrium of the solution, and so to cause new or increased local precipitation. In some cases the intersection is with a mineralized vein, and both veins along their intersection and for some distance back from it carry ore-shoots; in other cases, the intersection is with a barren, or unmineralized fracture, which appears equally likely to cause a segregation of values.

The relation between intersections and secondary ore-bodies is of perhaps even greater importance than in primary deposits.

Dr. Richard Beck[1] states that the localization of values is most marked where veins intersect at acute angles, because the intersecting vein walls are close together for longer distances and also because the mass of crushed and permeable rock is likely to be greater in acute than in more nearly rectangular intersections.

IN THE TORNADO-MOGUL SILICEOUS ORE-SHOOT, BLACK HILLS, SOUTH DAKOTA,[2] the large north-south ore-body is joined on the east by northeast ore-shoots; at the junctions the ores carry more gold and silver than do any of the shoots away from the intersections.

AT CRIPPLE CREEK, COLORADO,[3] the smaller ore-shoots are quite generally associated with intersections; the large ore-shoots appear to be independent of intersections.

[1] Beck-Weed, "Nature of Ore Deposits," p. 391.
[2] J. D. Irving, *Economic Geology*, Vol. III, p. 153.
[3] Waldemar Lindgren and F. L. Ransome, *P. P.* 54, U. S. G. S., p. 210–212.

IN THE GEORGETOWN QUADRANGLE, COLORADO,[1] a large proportion of the important ore-shoots occur at intersections with branch veins.

AT TONOPAH, NEVADA,[2] pre-mineral cross fractures and

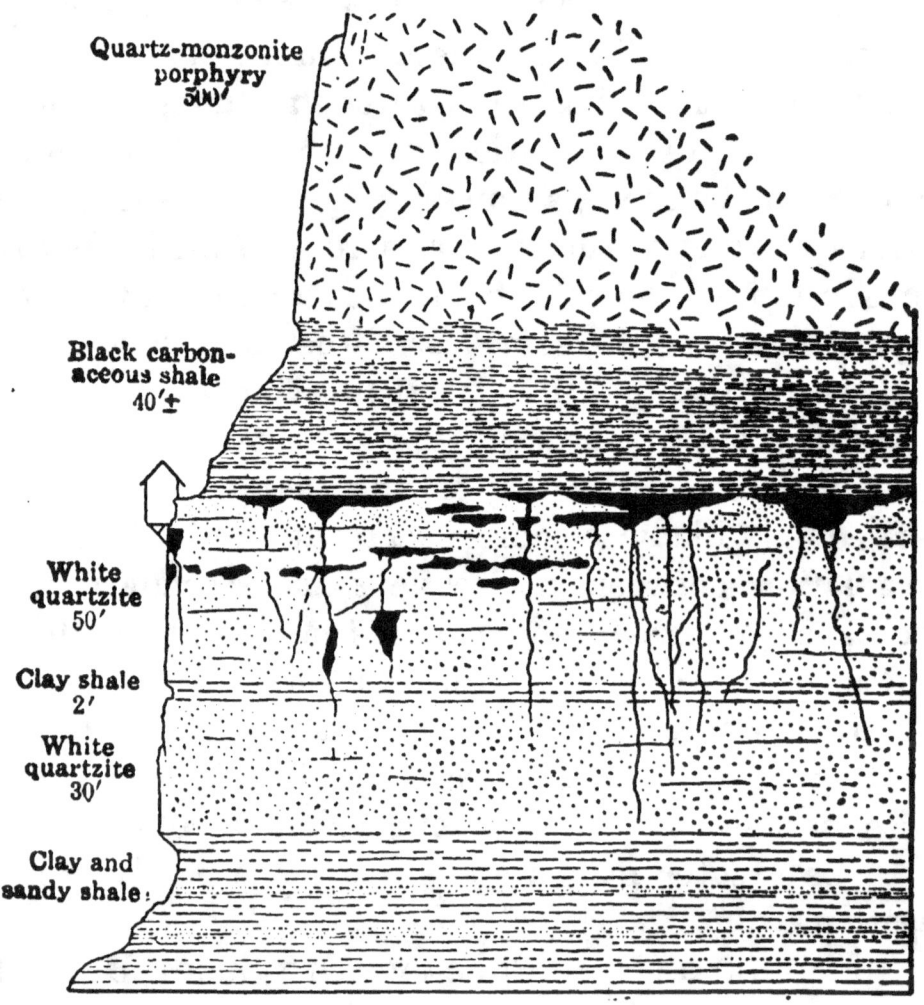

Quartz-monzonite
porphyry
500'

Black carbon-
aceous shale
40'±

White
quartzite
50'

Clay shale
2'

White
quartzite
30'

Clay and
sandy shale

FIG. 59.—Section of the rocks and diagram of the ore-bodies in the American Nettie mine, Ouray, Colorado, showing the localization of ore below an impervious stratum due to the impounding of solutions. *After Irving.*

cross walls appear to have been the cause of localizing values in ore-shoots, which, in general, pitch in a direction parallel to the intersections of these minor fissures with the main vein system. To quote Mr. Spurr: "To these cross walls, more

[1] J. E. Spurr, *P. P.* 63, U. S. G. S., p. 159.
[2] J. E. Spurr, *P. P.* 42, U. S. G. S., p. 85, and 119

or less pronounced, was due the division of the water circulation into columns of unequal importance, and the mineralization accomplished by these waters was correspondingly localized."

Ore-Shoots Due to the Impounding of Solutions.—A numerically

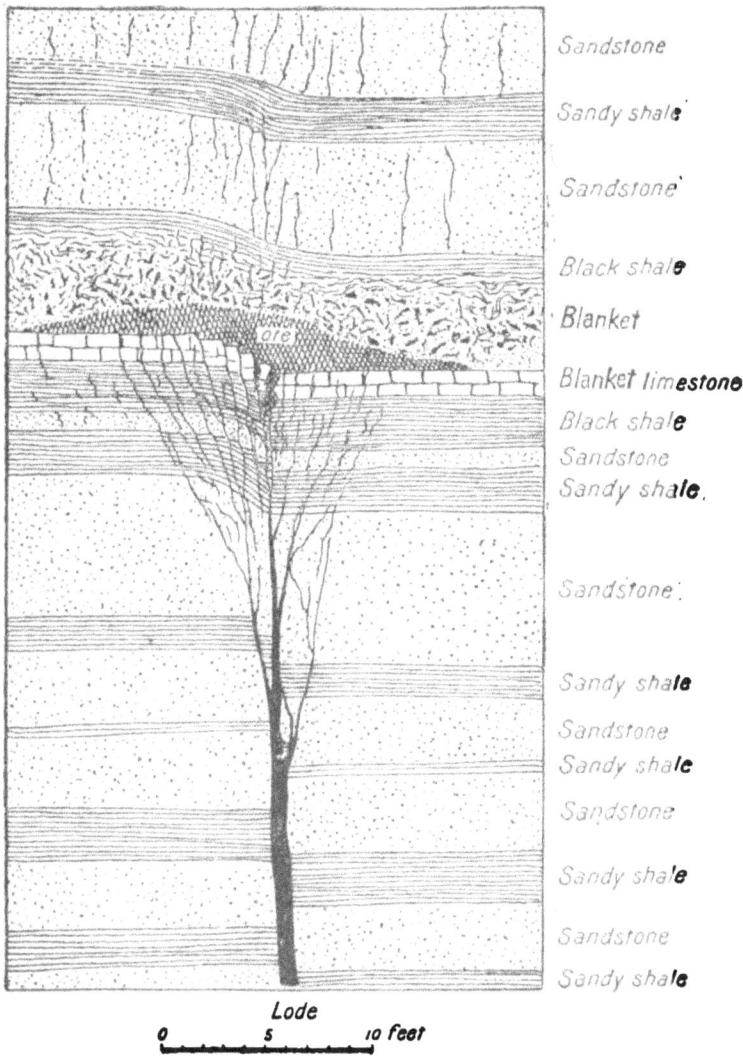

Fig. 60.—Diagrammatic section across a lode and ore-body formed beneath an impervious stratum, Rico, Colorado. *After Ransome.*

important type of ore-body appears to be the result of the impounding of mineralizing solutions beneath impervious cappings.

Fissures are formed only in rocks that are relatively rigid, and are likely to die out upon encountering a stratum or mass of rock that yields to the forces that elsewhere produce fissures. Most rocks are sufficiently rigid to permit the development of fractures, and the only rock of wide distribution that yields in preference to fracturing is shale; as shales are frequently intercalated with limestones and quartzites, ore-bodies due to the impounding of solutions are usually replacments of these rocks and the development of the ore-shoots may be considered as the effect of both structural and chemical causes.

It appears that where a mineralizing solution is impounded, or its circulation arrested, the increased time during which the mineralizing agents are in contact with the rock is an important factor in localizing deposition.   The fact that the farther upward passage of these solutions or mineralizing agents was almost, if not quite checked, indicates that the flow of solutions from below could not have continued long; large ore-bodies formed beneath impervious strata would seem to indicate, therefore, that the mineralizing agents possessed greater mobility than liquid solutions, and perhaps rose through such stagnant solutions to perform the work of ore deposition.

IN THE OURAY DISTRICT, COLORADO,[1] fissures of small displacement that are well developed in limestone and quartzite strata become lost upon meeting black shales, where they are represented by slight distortions of the beds only.   Beneath these impervious shale beds important replacement deposits have formed in the limestones and quartzites, apparently due to the impounding of rising solutions.

IN THE BLACK HILLS, SOUTH DAKOTA,[2] the vertical fissures pass through limestone into impervious shales: the ore-bodies, which are replacements of the limestone, are widest just beneath the shales, and narrowest in their downward extension, apparently from the impounding of rising waters.   In the great Tornado-Mogul ore-shoot this occurs on a large scale.   Here a phonolite dike forms the west wall of the ore-body, and not only has the upward progress of the solutions been prevented by overlying shales, but they have been confined laterally also by the porphyry wall.

[1] J. D. Irving, *Bull.* 260, U. S. G. S., pp. 58, 62.
[2] J. D. Irving, *Economic Geology*, Vol. III, p. 150, and *P. P.* 26, U. S. G. S.

IN GRANT COUNTY, NEW MEXICO,[1] silver, and silver-gold ores occur in limestone immediately below strata of shale, which, by impounding solutions, have had a distinct effect in the localization

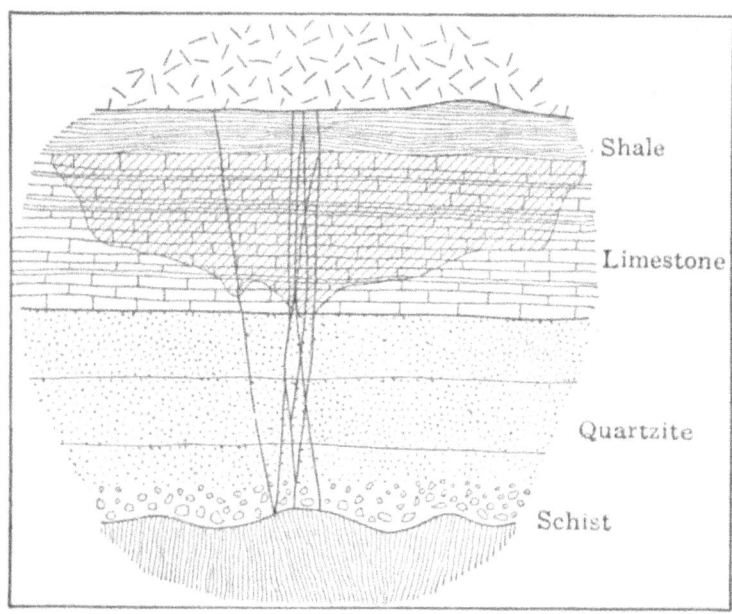

FIG. 61.—Sketch showing narrow mineralizing fissures and the replacement of limestone by ore beneath a stratum of impervious shale, Alameda mine, Portland, South Dakota. *After Irving.*

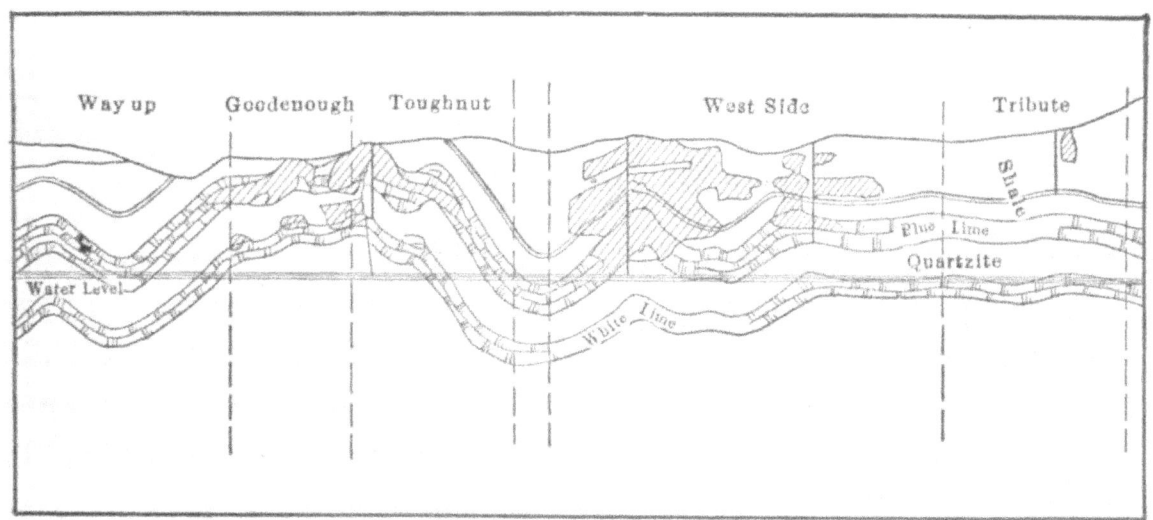

FIG. 62.—Section along the West Side Vein, Tombstone, Arizona, showing the ore-bodies developed at the tops of anticlines. *After Church.*

of the ore-bodies at Kingston, Lake Valley, Hermosa, and other camps.

[1] C. H. Gordon, *P. P.* 68, U. S. G. S., p. 269.

Ore-Shoots at the Tops of Anticlines, or "Saddle Reefs."—In some districts the ore-deposits occur persistently at the tops of anticlinal folds. In many cases this appears to be the result of local impounding of ascending solutions at the tops of anticlines that are capped by strata of impervious rock. Fractures and open spaces form in the convex side of folded strata somewhat before the production of a continuous fissure, and the largest

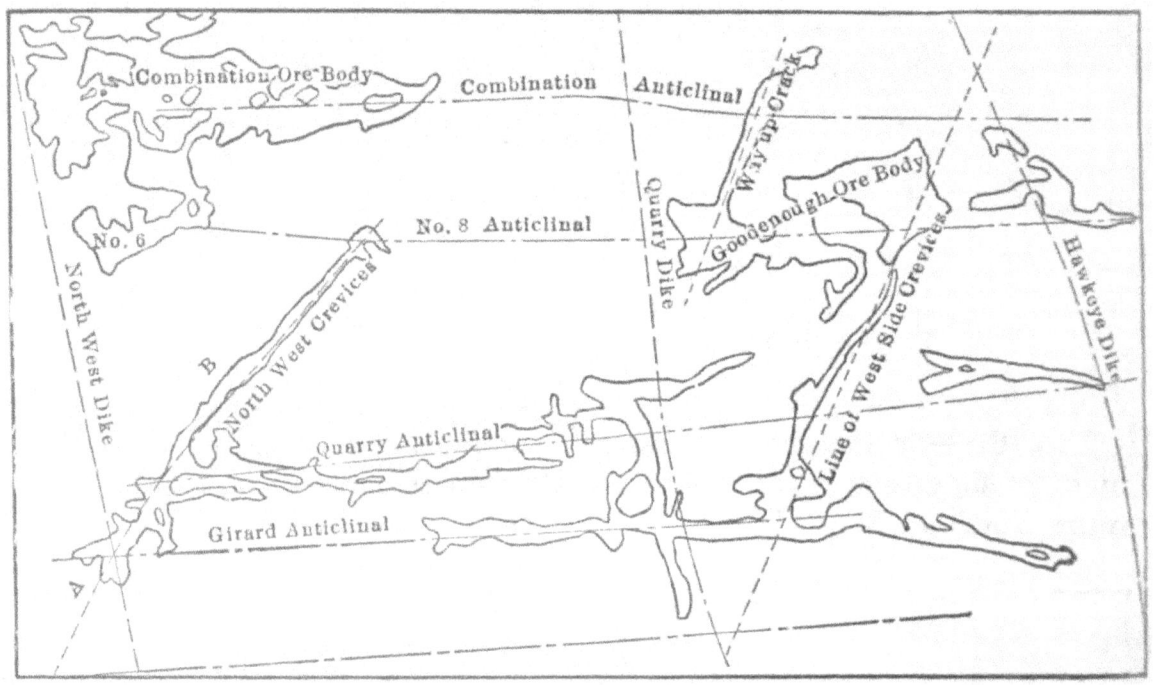

Fig. 63.—Plan of ore-bodies, Tombstone, Arizona, showing their relation to anticlinal folds. *After Church.*

open spaces along fissures through folded rocks are often found at such points; these open spaces appear to be favorable to the segregation of ore minerals. Rising solutions, therefore, are likely to form ore-bodies at the tops of anticlines. The same relation appears to hold true for the localization of secondary ores in synclinal troughs by descending solutions.

At Tombstone, Arizona,[1] the deposits lie in the anticlines, in some cases on the flank, in other cases at the apex, but the synclines are barren. Ore-shoots in the veins are associated with anticlinal folds of the wall rock.

[1] John A. Church, *Trans.* A. I. M. E., XXXIII, p. 14.

At Bendigo, Australia,[1] quartz veins traverse highly folded shales and sandstones. The ore-deposits are local developments at the tops of the anticlines, and in typical cases a series of these so-called "saddles" are superposed; they are usually connected by stringers with underlying and overlying saddles, or with the mineralizing fissure.

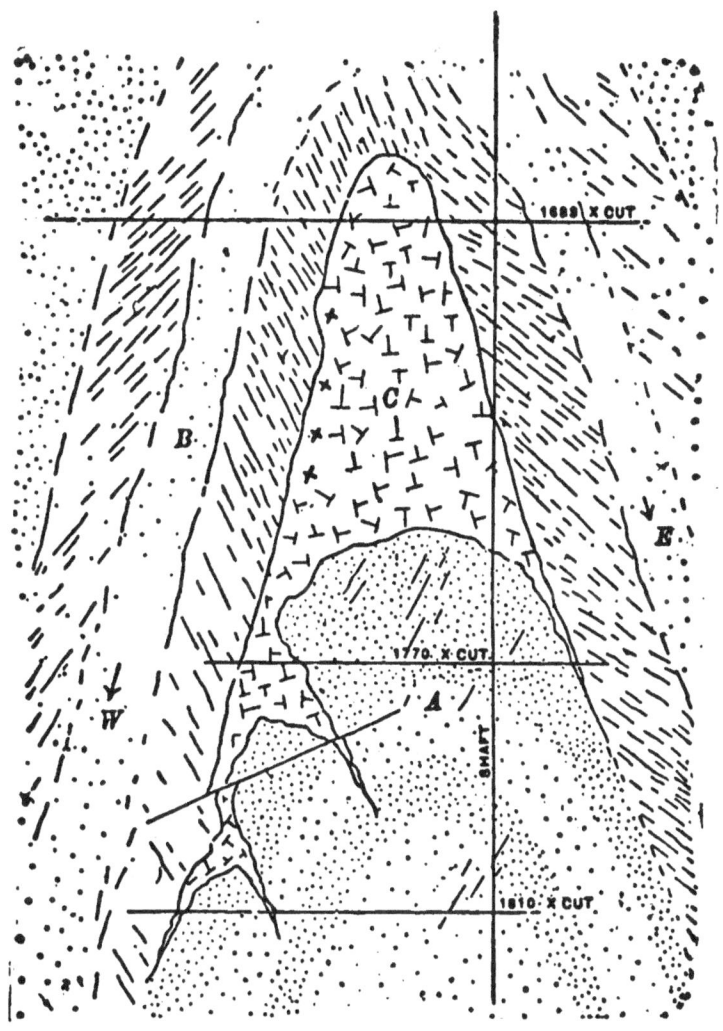

Fig. 64.—Section through a saddle reef, Bendigo, Victoria. *A*, Sandstone; *B*, shaly sandstone; *C*, quartz-ore. *After Rickard.*

In the Nova Scotian Gold Fields.—Mr. E. R. Faribault[2] has described a condition similar to that at Bendigo, Australia. In the example given, of 21 arches of strata on the anticlines, 20 contain gold ore. The cause of the localization appears to

[1] T. A. Rickard, *Trans.*, A. I. M. E., XX, p. 480.
[2] *Trans.*, Can. M. I., 1899; Beck-Weed, p. 472.

have been opening or parting due to slipping, as shown by slickensides, along the contact planes between quartzite and slate. That the parting and deposition was gradual and progressive is shown by the crustified structure of the ore.

AT THE ELKHORN MINE, ELKHORN, MONTANA,[1] the principal deposit occurs as a saddle-shaped mass along the axis of a steeply pitching anticlinal fold. The ore occurs at a contact between altered shale and a massive crystalline limestone. The bedding planes are ore bearing only where the general dip of the rocks is disturbed by flexures. In addition to the ore formed along the contact, a number of very large ore-bodies have been found at some little distance from it in the dolomite, but always in the same structural position.

The Chemical Influence of Wall Rocks on Ore-Shoots.—The chemical effects of different wall rocks upon the segregation of minerals into ore-shoots is well established by many clear examples.

A vein that indicates by a well-developed banded structure or crustification its origin through the filling of open spaces is unlikely to vary in mineral content in passing from one country rock into another, except as the structural character of the original fissure varies in the different rocks. In veins that were formed by a replacement of the walls of their fissures, however, the mineral content is likely to vary markedly in different rocks, in proportion to the precipitative action, permeability, and solubility of the several rocks encountered.

As the chemical activity of any rock mass must vary with the character of different mineralizing solutions, being greater or less according to the kinds and amounts of the dissolved materials, no general rule may be formulated to apply to all veins, nor to any given type of country rock. In general, however, where a vein passes from a massive and insoluble rock to one that is relatively pervious, soluble, and replaceable, the probabilities are that the usual vein filling will expand and a relatively large replacement body will be found in the favorable rock. Such

[1] W. H. Weed, *Trans.*, A. I. M. E., XXXI, p. 647.

enlargements frequently extend outward for considerable distances from the mineralizing vein, which may indeed be an insignificant and unmineralized fissure, known only through its effect in having transported solutions to a rock horizon suitable for the precipitation of their mineral content.

A distinction is difficult between the practically very different but genetically similar types whose extremes are represented by

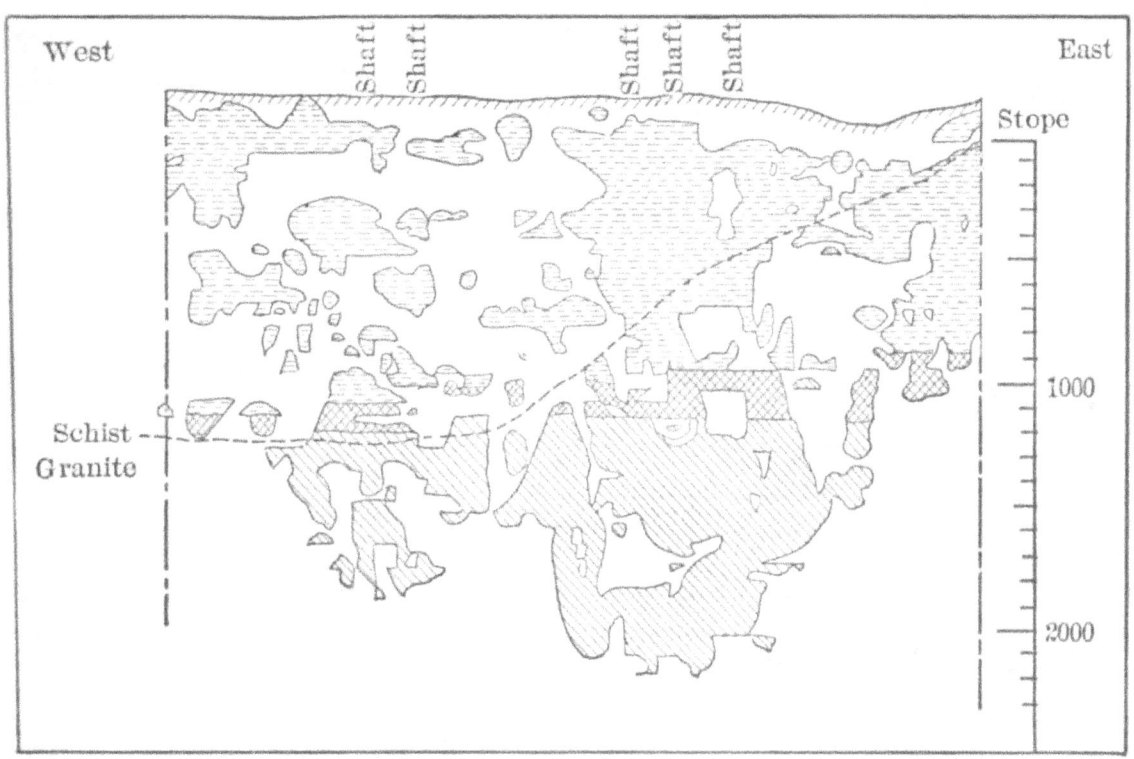

FIG. 65.—Section through the Dolcoath mine, Cornwall, showing the copper stopes where the vein is in schists, and the tin stopes where the vein traverses granite. *After LeNeve Foster.*

replacement deposits on one hand, and by typical veins whose ore-shoots are the result of chemical action of the wall rock on the other.

The relative precipitative action of the various rocks upon mineralizing solutions may not be stated definitely, but there are certain general relations that may be considered well demonstrated in spite of exceptions. The sedimentary rocks are, in general, more active chemical precipitants and more easily replaceable than are the igneous rocks, although this is not invariably the case. Limestones are among the most chemically active rocks in precipitating minerals from solutions, and are most easily replaceable by mineralizing solutions of usual composition. Next to limestones come calcareous shales, sandstones, quartzites and conglomerates.

In each case the chemical and physical character of the rock and of the mineralizing solution determines the most favorable locus for deposition, and local data, as usual, is of more value in predicting the most favorable horizon for exploration than is the best authenticated general data. Where several beds of the

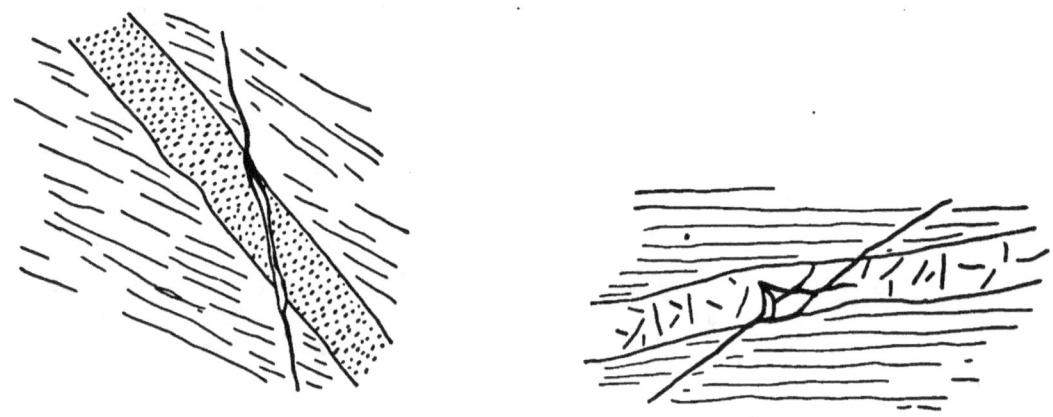

Fig. 66.—Sections showing the behavior of veins in passing through porphyry dikes, Cornwall, England; the values are segregated where the porphyry forms the walls of the veins. *After De la Beche.*

same sediments are present, it is usual that one particular bed has been the most active chemically, and carries the ore-deposits to the perhaps complete exclusion of the other beds of the series. Carbon is an active precipitant of minerals from solutions. The rock or bed that will be most active chemically in precipitating minerals and will contain the richest or largest ore-bodies may occasionally be predicted from its being known to contain carbonaceous matter. Most sedimentary beds contain iron, and not infrequently this combines with sulphur contained in carbonaceous material to form pyrite, which is an active precipitant of other sulphides and of gold. This condition is typical of shales. The influence of carbon, contained in wall rocks, upon the precipitation of ores, while undoubtedly the controlling factor in some districts, has probably been over-rated; the amount and distribution of carbonaceous matter is wholly incapable of explaining the occurrences of many large replacement deposits; its chief function may be, perhaps, to start a precipitation which is carried on through mass action in the easily replaceable rock.

The selective precipitative action of different igneous rocks is also well established, but although the different effects of various rocks in segregating values may be apparent, the differing characteristics that produce these effects are not usually notice-able, and in advance of local information, no one rock may be considered more favorable for ore deposition than another. Many veins pass through several igneous rocks without apparent change in value or mineral content.

In the exploration of a vein in depth a cessation of payable ore should not immediately be assigned to primary impoverishment due to depth alone, for it has been demonstrated that such impoverishment frequently, or even usually, takes place gradu-ally and over great vertical distances. In cases where the segre-gation of ore into shoots may be referred to the precipitative action of wall rocks, the possibility of a reccurrence of the favorable wall rock at greater depths at once suggests itself. If the exposed ore is due to geological factors unlikely to recur in depth, then deeper exploration is unwarranted; if the geo-logical conditions appear to be as good in depth as in the horizon of known productivity, then deeper exploration for primary ore may be justified.

In the secondary concentration of metals by surface agencies the character of the country rock is of the greatest importance, some rocks actively precipitating metals and not permitting the migration necessary to form enrichments.

Where it is seen that a vein is to pass into a different rock, the change will be awaited with anxiety unless it is known that the vein has been formed through the filling of open spaces, when a change is much less likely than in a vein formed partly by re-placement. It is seldom that such a change in wall rocks may be expected to accompany better ore, unless the expectation is based upon abundant local-evidence.

At Neihart, Montana,[1] the primary ore consists of galena zincblende and pyrite in a gangue of carbonates of lime, iron and manganese; the veins carry ore-shoots between walls of pink or white feldspathic gneiss, but are barren where they pass through dark colored gneisses, amphibolite and diorite. The mineralizing solutions appear to have reacted with the feldspars and not with the ferro-magnesian silicates.

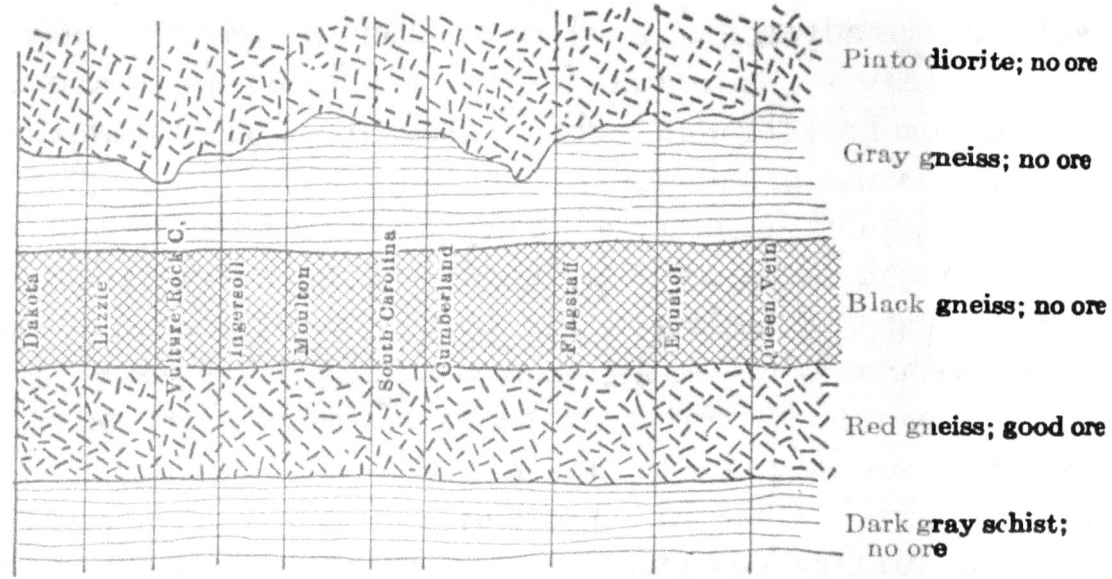

Pinto diorite; no ore

Gray gneiss; no ore

Black gneiss; no ore

Red gneiss; good ore

Dark gray schist; no ore

Fig. 67.—Diagram (plan) showing the rocks traversed by the Neihart veins and their ore content in the various rocks. *After Weed.*

At Butte, Montana,[2] a series of copper veins intersect quartz-monzonite, aplite, and quartz-porphyry. The veins were formed through replacement of the wall rocks, which are much altered; in the monzonite they are commonly rich in copper; in the aplite they are equally wide and strong, but are lean, being composed chiefly of quartz with comparatively little pyrite and copper. In the porphyry the veins are narrow and lean. The pyrite contained in the veins where they pass through aplite and the quartz-porphyry is noticeably poor in copper.

[1] W. H. Weed, *Trans.*, A. I. M. E., XXXI, p. 645.
[2] W. H. Weed, *Trans.*, A. I. M. E., XXXI, p. 642.

In this district the ore-shoots favor the more basic rock. The silver veins at Butte exhibit a marked crustification, and are clearly the result of the filling of open fissures. They traverse both the monzonite and the aplite, but there is no perceptible difference in their character or tenor where they pass through the different rocks.

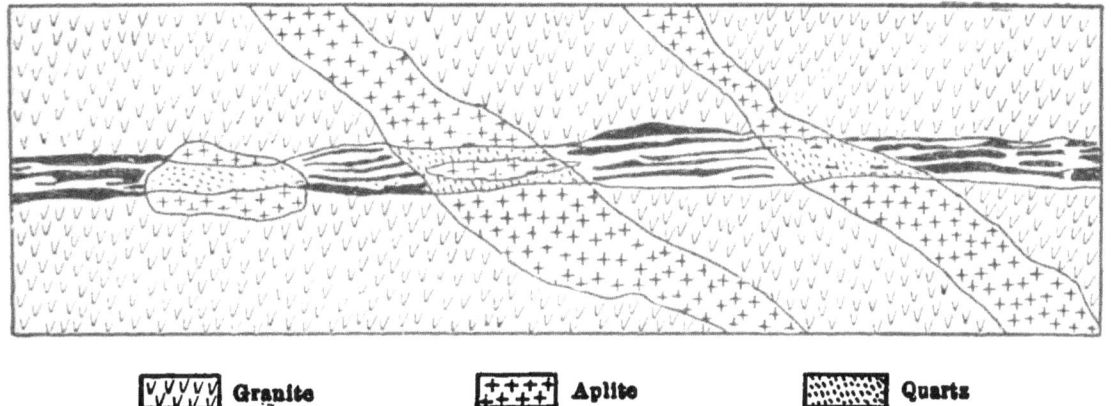

⟦V V V V V⟧ **Granite**   ⟦+ + + + +⟧ **Aplite**   ⟦▒▒▒▒▒⟧ **Quarts**

Fig. 68.—Ideal plan of a copper vein, Butte, Montana, showing the impoverishment where aplite forms the walls of the vein; the ore is indicated in black. *After Weed.*

At Gympie, Eastern Australia,[1] the precipitative influence of a slate band is so marked that it is followed in exploration instead of the veins. The main slate band is 200 ft. in thickness, and the several veins carry ore where they pass through it, but are elsewhere unpayable.

At Aspen, Colorado,[2] the mineralizing solutions appear to have risen through the underlying siliceous formations (granite and quartzite) with but little precipitation of ores, and through the overlying dolomitic and calcareous formations with much precipitation, and finally, upon encountering the overlying shales and the prophyry sheet that was intruded at their base, the solutions were impounded, and the shales reacted powerfully with the solutions for the production of large deposits.

At Ballarat, Australia,[3] the prevailing country rocks are

[1] J. H. Curle, "Gold Mines of the World," p. 188.
[2] J. E. Spurr, *Economic Geology*, IV, p. 310.
[3] T. A. Rickard, *Trans.*, A. I. M. E., XXX, p. 1010.

slates and sandstones that carry as intercalated members a series of black slate bands, which contain carbonaceous matter and pyrite. These bands are persistent over long distances and while typically very thin, being from a fraction of an inch to a few inches in thickness only, they have had a remarkable effect

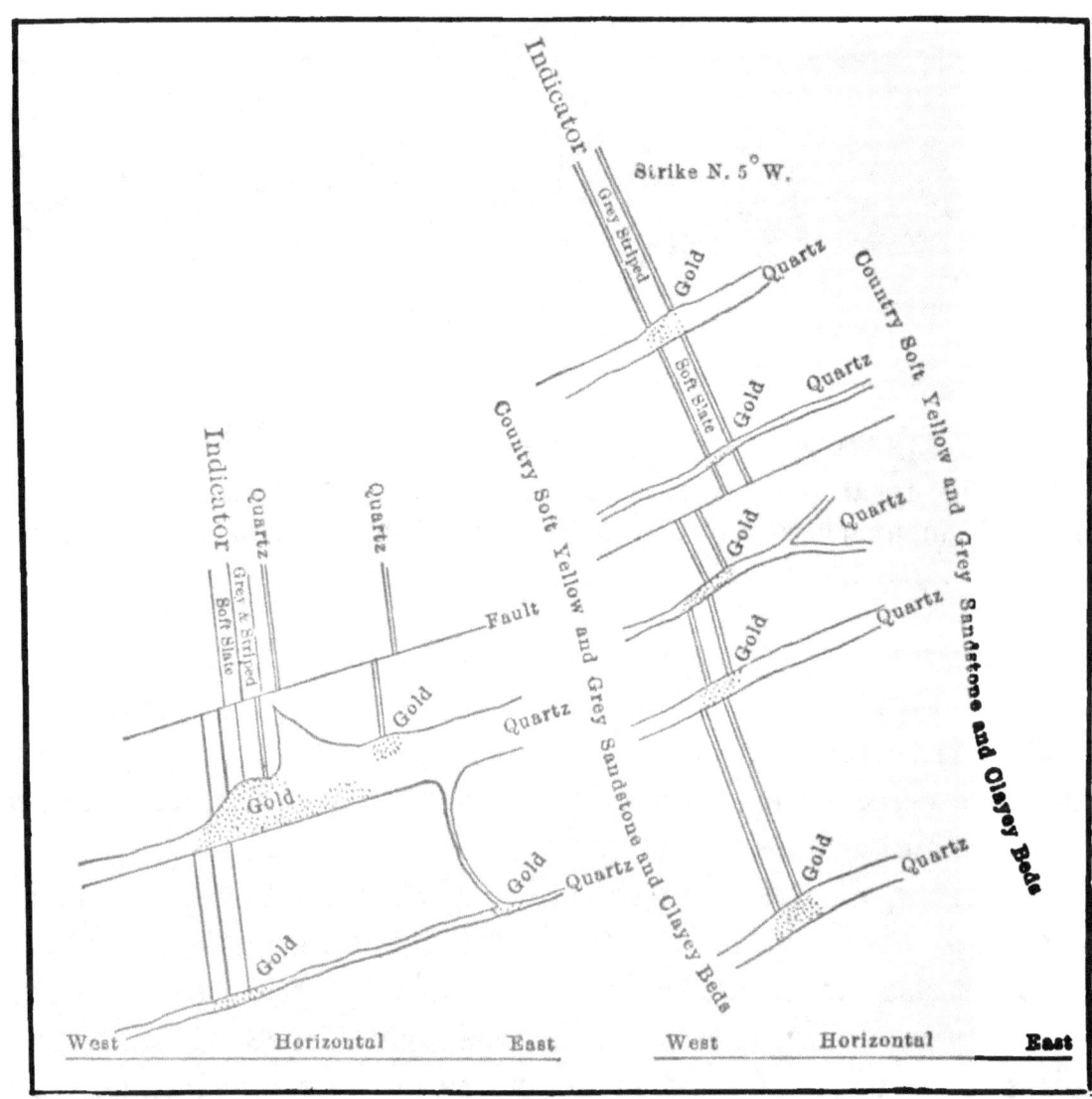

Fig. 69.—Plans of the "Indicator" slate seam near Ballarat, Australia, showing the segregation of gold in the intersecting quartz veins. *After Dunn.*

in segregating the gold content of a series of quartz veins that intersect them. The chief of these slate seams is known as the "Indicator" and along the intersections of the quartz veins with it, as well as with the other seams, the ore-shoots have formed.

At San Javier and Tecoripa, Sonora, Mexico, an extreme case of the precipitative action of carbon upon mineralizing solutions is evident. Here the solutions have followed steeply dipping and much altered coal seams; the carbonaceous matter has in great part been altered to graphite, with which are associated the rich silver ores.

In the San Juan Mountains, Colorado,[1] the distribution of the primary ore in veins that cut a series of volcanic beds appears to be due to the greater precipitative effect of certain beds over others. The veins are commonly low in grade in the rhyolitic layers, and of good grade, frequently of high grade, in the andesitic layers. It is said that in the San Juan andesitic breccias the value of the ore varies from layer to layer. Ore deposition here appears to have been greater in the basic than in the siliceous rocks.

In the Plateau Region of Arizona and New Mexico, there are many districts where tree trunks contained in red Triassic sandstones have by replacement been transformed wholly or in part to chalcocite, which appears in this connection to be a primary mineral. The quantity of ore being limited to the carbonaceous material, which was scanty in amount and erratic in distribution, these tree trunk deposits have yielded small commercial returns.

Near San Bernardo, Sonora, Mexico, a small quantity of high-grade ore was mined where a tree trunk had been replaced by chalcocite and silver glance. The whole tree was extracted, including several branches, when the "mine" came to an end.

Ore-Shoots in Veins of Deep-seated Origin.—In many veins of deep-seated origin the ore occurs in well-defined shoots which, apparently, are not connected with any of the factors discussed in the preceding paragraphs; to this class belong some of the most important ore-shoots known.

Inasmuch as a localization of values must be assigned to a local difference or cause, these shoots are probably best referred to either mass action, or to a change in the predominant circula-

[1] C. W. Purington, *Economic Geology*, I, p. 133.

tion channels at different times in the formation of the vein or deposit. In the former case, a precipitation of a metal once started, by any cause—and all precipitations must have a local start— is continued progressively, the mineral formed attracting to itself additions of the same kind from the solutions until the result is a localization of values, or an ore-shoot. In the latter case, it is difficult to conceive that a fissure at all times permits the passage of solutions over its entire length with equal facility; therefore, at any given time certain parts of a fissure probably constitute the main channels for the passage of solutions and so give rise to the localization of values in ore-shoots. It is probable that veins, constituting lines of weakness, have been opened and reopened many times during ore deposition; the ore-shoots, therefore, may be considered to represent the circulation channels at the time of passage of the richest solutions.

Many ore-shoots are of lenticular shape, being widest at the center and decreasing in thickness toward their peripheries, where they die out in stringers and fade into barren material. The study of fissures themselves indicates for them a similar form, inasmuch as they are strongest over their central portions, and die out at their ends in stringers or fade out in the rock where the stresses that formed them were absorbed without fracture. This type of shoot may, therefore, be referred to a reopening of the vein in a manner similar to that of the formation of fissures through rocks.

Certain ore-shoots of great depth as compared to width may be due to incipient folds on the plane of the vein, such lines of weakness in the vein as were most nearly vertical affording the most favorable channels for rising solutions.

CHAPTER VII

THE PRIMARY ALTERATION OF WALL ROCKS

Metamorphic Processes.—The term metamorphism as commonly used means any change in a rock either in form or composition, from whatever cause. By metasomatism is meant a metamorphism that involves a change in the chemical composition of a rock by the addition or subtraction of substance.[1]

Certain types of metamorphism are due to the action of mineralizing solutions, and are closely connected with ore deposition; they form, therefore, valuable guides in the exploration for ore-deposits. Other types of metamorphism are due to regional processes, and are not connected with ore deposition.

Dynamo-regional Metamorphism.—By dynamo-regional metamorphism is meant a primary alteration of rocks over large areas under conditions of pressure, flowage, and heat. This process produces schists and gneisses from either sediments or igneous rocks. It is thought that there is little migration of minerals or of elements in dynamo-regional metamorphism, and that the changes are due chiefly to recrystallization and rearrangement of original minerals in bands with their major axes oriented normal to the direction of greatest pressure. Under these conditions limestones are commonly altered to marble, and sandstones to quartzite, basic igneous rocks to greenstone or serpentine, while rocks of normal or acid composition commonly yield the usual types of schist or gneiss.

Dynamo-regional metamorphism, extending over large areas, has no connection with mineralizing processes, and is, therefore, no guide to ore-deposits; while this process has certain features, such as characteristic mineral associations, in common with contact metamorphism, the results of the two processes may usually be distinguished without difficulty.

[1] Waldemar Lindgren, *Trans.*, A. I. M. E., XXX, p. 580.

Contact Metamorphism.—By contact metamorphism is meant the change in structure and composition of the enclosing rocks in the immediate vicinity of and due to accession of substance,[1] from igneous intrusions.   The rocks most affected by this process are limestones, shales, and sandstones, which are altered respectively to marble, hornfels, and quartzites with the further addition of new minerals; igneous rocks are affected to a less extent by contact metamorphism, having themselves been formed under igneous conditions.   The characteristic contact metamorphic minerals are garnet, epidote, wollastonite, pyroxene, amphibole, magnetite, quartz and sulphides; the unique feature of contact metamorphic deposits is the primary association of oxides with sulphides.[2]   While contact metamorphism is frequently associated with ore deposition, the development of a large zone of contact minerals does not presuppose or indicate the existence of ore-deposits.

The distance to which contact metamorphism effects the altered rocks varies greatly in different occurrences, and also in different beds of a sedimentary series, certain beds being altered for long distances from the contact while others may persist unchanged up to within a short distance of the intrusive.   Contact metamorphic minerals and products are commonly tough and resistant to erosion, and are, therefore, likely to form prominent outcrops.

Hydrothermal Metamorphism.[3]—By hydrothermal metamorphism is meant the alteration of rocks by hot ascending solutions; it is distinctly a metasomatic process and is a frequent accompaniment of ore deposition.

The walls of veins are commonly fractured or strained by the stresses that produced their fissures, especially the hanging walls, and in this way ready access is afforded to the solutions that accomplish their alteration.   The alteration is usually intense

---

[1] Waldemar Lindgren, *P. P.* 43, U. S. G. S., p. 124.

[2] Waldemar Lindgren, *Trans.*, A. I. M. E., XXXI, p. 227.

[3] The present knowledge of metasomatic replacement and of hydrothermal metamorphism is chiefly due to Mr. Waldemar Lindgren.

immediately along the veins, but is likely to diminish rapidly in degree with distance from them. It appears that the walls of a vein tend to keep within them the heavy bases, and gangue minerals, but permit the passage of the depleted solutions that effect the rock alteration. Not infrequently, disseminated sulphides are found in the altered wall rocks, although these minerals commonly carry low values in gold and silver as compared with the sulphides deposited within the veins. Pyrite is the most widely disseminated sulphide in wall rocks, and is frequently the result there of the alteration of the rock minerals themselves, in which case it does not carry values in the precious metals. Zincblende and galena are more rarely found as disseminations.

The results of hydrothermal metamorphism are frequently valuable guides in the search for ore-deposits, especially where they are confined to the vicinity of the veins or stocks. In many cases, however, and especially where mineralization and attendant primary rock alteration have taken place at slight depth below the surface, the porosity and shattered condition of the wall rocks has permitted a wide distribution of the altering waters, and the altered areas are likely to be large, and so to lose their value as aids in the search for ore-bodies.

In many parts of the arid regions of the western United States and Mexico there are large areas of brilliant red hills, the rocks of which have suffered hydrothermal metamorphism; the iron mineralization of these red hills is likely to be more apparent than real, the surface being stained a deep red, yellow or brown, but upon fresh fractures the rock is seen to carry little iron, present in most cases as pyrite, resulting from the alteration of the rock minerals.

The alteration of wall rocks is likely to be less along deep-seated veins than in deposits of shallower origin; it is also likely to reach a less development where it accompanies deposits resulting from the filling of open spaces than in connection with replacement deposits.

The most common effect of intense alteration immediately

along veins is sericitization and the introduction of pyrite, which alteration is likely to grade into propylitization with greater distance from the veins, the latter process being the less intense and more widespread; in basic rocks, propylitization is likely to persist close up to the veins, there to give way perhaps to sericitization. While the degree and extent of alteration by one set of solutions is likely to vary markedly in different rocks, it not infrequently happens that the final products of alteration and recrystallization are similar, and it becomes a matter of difficulty to distinguish one rock from another. The chief chemical effects of hydrothermal metamorphism are an increase in potash with attendant loss in soda, silicification, and the introduction of sulphides, chiefly pyrite.

The effects of hydrothermal metamorphism may usually be distinguished readily from the effects of hydro-metamorphism, or the alteration accomplished by surface waters: the latter alteration is rather more likely to be uniform, and is essentially a process of oxidation, hydration, and solution, while the former is characterized by sericitization and silicification. Kaolin is the chief guide in making this distinction, being a typically surface product, often formed from sericite as well as directly from feldspars through the action of sulphuric acid solutions set free by the oxidation of pyritic sulphides.

Propylitization.—By propylitization is meant a hydrothermal metamorphism that involves the development of chlorite, epidote and pyrite from the dark rock-making silicates, and of quartz, calcite, and epidote from the feldspars. Rocks that have been subjected to propylitic alteration are commonly greenish-gray in color and carry bright green stains of epidote; the barren pyrite developed is present in well-developed crystals, and upon oxidation is likely to color the surface red, brown or yellow.

This type of rock alteration is frequently present over large areas in mineralized districts, and is rarely a guide to ore-deposits, except as indicating in general that a district has been subjected to metamorphic action.

Propylitization is thought to have taken place at no great

depth beneath the surface, where extensive fracturing and jointing permitted easy access by solutions, and although associated in a general way with ore deposition, is rarely to be referred to any particular fissure or series of fissures. In rocks that have suffered propylitic alteration the feldspars have usually become dull, if not more completely altered, but the principal physical features of the rocks, such as texture, are preserved. Propylitic alteration is most prominently developed in rocks of intermediate or basic composition.

Sericitization.—By sericitization is meant a hydrothermal metamorphism that results in an almost complete loss of soda and a large gain in potash, silica, and commonly also of pyrite, with occasionally lesser quantites of carbonic acid and fluorine. The typical result of complete sericitization is a finely granular aggregate of sericite, quartz, pyrite and calcite; after thorough sericitization it is difficult to distinguish between rocks that were originally quite dissimilar.

Sericitization is a common form of hydrothermal metamorphism, and is an intimate accompaniment of the mineralization of most ore-deposits in igneous rocks; the alteration is commonly most intense immediately along the veins, and decreases with distance from them, sometimes passing into a propylitic alteration, which latter transition is likely to take place within a short distance of the veins in basic rocks; sericitization not infrequently persists over large areas in rocks of intermediate or acid composition.

Sericitized rocks are commonly white to light yellow in color, and are usually soft; the minute scales or plates of sericite are usually distinguishible in hand specimens and often form a means of distinguishing a sericitized from a kaolinized rock; the two minerals are likely to occur together, however, the latter being formed from the former as well as from feldspars by the action of surface waters carrying sulphuric acid. The pyrite introduced in sericitization is sometimes accompanied by minute quantities of chalcopyrite and zincblende, less frequently by galena or arsenopyrite.

AT BISBEE, ARIZONA.[1]—The porphyry of Sacramento Hill and the adjacent schists have undergone sericitic alteration. The results of alteration of both rocks are similar. The original quartz has recrystallized into quartz aggregates, the feldspars have become areas of sericite, quartz and pyrite, with occasionally a little epidote, chlorite, zircon and rutile. Under the action of surface agencies the pyrite has oxidized, and much of the sericite has been altered to kaolin. The surface rock consists chiefly of an aggregate of quartz, stained and streaked by iron, and containing nests of sericite and kaolin at the surface. Much of the kaolin has weathered out, leaving cavities. The outcrops are brownish-red in color, but the rock in most exposures is white on fresh fractures.

Alunitic Alteration.[2]—The occurrence of alunite as a primary mineral has been established at Goldfield, Nevada. It is here associated with gold, sulphides, tellurides, sulphantimonites, and kaolin, as a primary result of hydrothermal metamorphism. Kaolin is elsewhere regarded as a typical product of the action of surface waters. The alunitic alteration at Goldfield is widespread, and is intimately associated with ore deposition. Alunite occurs massive, as a crystalline constituent of the altered rocks in the ore-bearing areas, and intercrystallized with pyrite, gold, tellurides and other minerals in the ores.

Alteration to Greisen Along Tin Veins.—The wall rocks of tin veins are characteristically altered by hydrothermal metamorphism to greisen, the alteration consisting of the replacement of feldspars and of biotite by quartz, lepidolite, topaz, tourmaline, and fluorite, commonly associated with cassiterite, wolframite, chalcopyrite and arsenopyrite. The altered rock in this case frequently constitutes payable ore. Greisen is commonly referred to as altered granite. Where the altered walls are other rocks, the same minerals are likely to be developed.

Silicification.—Silicification of wall rocks by hydrothermal metamorphism is a common form of primary alteration that is

[1] F. L. Ransome, *P. P.* 21, U. S. G. S., p. 150.
[2] F. L. Ransome, *P. P.* 66, U. S. G. S., p. 192.

frequently associated with ore deposition. The tendency toward silicification is probably greater in acid than in basic rocks, but the process is common also in limestones, where it is likely to

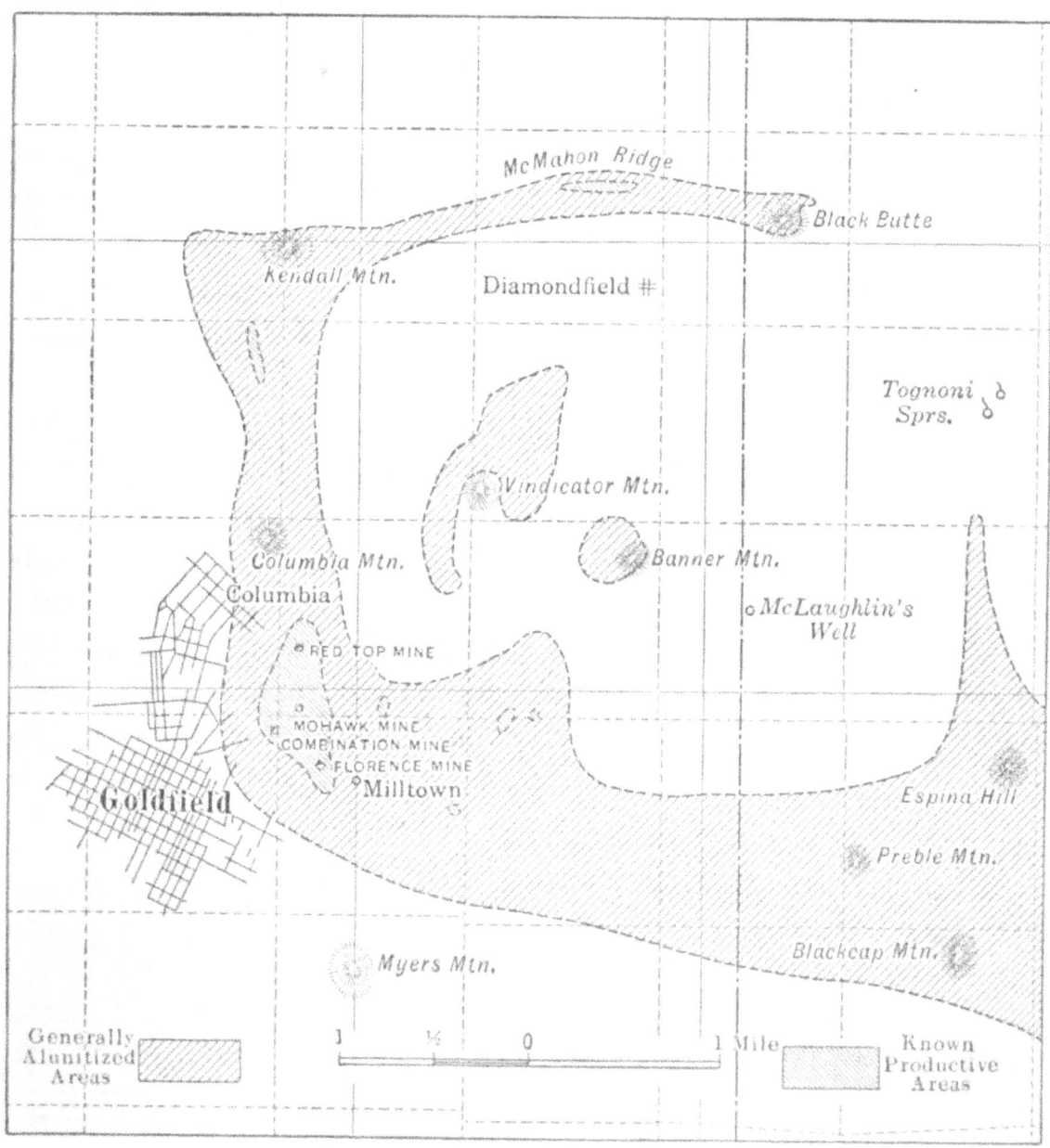

FIG. 70.—Diagram showing the alunitized areas and the known productive areas at Goldfield, Nevada. *After Ransome.*

form a reticulated structure, siliceous seams enclosing partially altered or fresh fragments. Quartz formed in this manner is commonly of the cherty variety; the replacement of calcite by silica often reproduces closely the original structure, where the difference in hardness between the altered and the unaltered parts offers the best means of distinction.

Silicification frequently accompanies other forms of hydrothermal metamorphism, where it is likely to represent the extreme of alteration, and thus serve as a valuable guide in a search for ore-deposits.

Marmorization.—The alteration of limestone to marble is frequently accomplished by contact and by hydrothermal as well as by regional metamorphism. A large, thoroughly marmorized area is probably without significance of mineralization, but local or irregular marmorization along contacts with a mineralizing intrusive, or along fissures, is not infrequently associated with ore deposition.

Dolomitization.—The replacement of a part of the calcium in limestone by magnesium is called dolomitization, and the resulting rock, dolomite. This change is commonly effected over large areas, and has no connection with ore deposition. Dolomitization is occasionally, however, of local occurrence, and as such may in some cases be the accompaniment of ore deposition. This alteration is accompanied by an important shrinkage in volume, and may thus indirectly play a part in ore deposition.

CHAPTER VIII

ALTERATIONS BY SURFACE AGENCIES

By hydrometamorphism is meant the alteration of rocks, ores and minerals by atmospheric waters. In its broadest sense, it includes the varied processes of weathering, oxidation, hydration, the leaching of rocks and ores, and the solution, migration and enrichment of metals.

Rain water, as it strikes the earth, is practically pure water, except for small quantities of dissolved atmospheric gases; upon coming in contact with ore minerals, especially sulphides, however, it is transformed into solutions of great chemical activity, which attack, transform and rearrange the minerals of ore-bodies, and so effect enrichments of the greatest economic importance. Primary deposits so low in grade as to be without commercial value are frequently thus transformed into deposits of commercial importance. It is necessary in the examination of any ore-body, therefore, to determine whether the valuable ore is of primary or of secondary origin; if primary, its values may be expected to continue indefinitely in depth to the zone of primary impoverishment; if secondary, the values are known to be controlled by surface agencies, and the valuable ore may be expected to continue to such depths only as have been reached by surface waters.

The Decomposition and Weathering of Rocks.—The decomposition and weathering of rocks, unlike the changes undergone by ore-bodies, are accomplished by water carrying as dissolved constituents chiefly atmospheric oxygen and carbon dioxide, together with vegetable acids. The chemical changes wrought are principally hydration, oxidation and the formation of carbonates. The tendency is toward the solution of the more easily dissolved minerals, which is attended by the formation and a surface concentration of the more resistant minerals, among which are quartz, aluminous clays and limonite; complex minerals are thus broken down into simpler constituents more resistant to weathering, which remain at the surface.

The weathering of rocks is characteristically regular over large areas, and is rarely mistaken for the results of hydrothermal alteration or kaolinization. The residual minerals from rock decomposition are commonly soft, and are readily carried away by erosion, which is likely constantly to present fresh rock surfaces to the attack of the chemical alteration above outlined. Even in regions of heavy rainfall, however, a combination of slight slopes and abundant and active vegetable acids may result in deep accumulations of residual minerals, as frequently occurs in tropical countries.

Weathering attended by slight or partial decomposition results chiefly in the staining of the surface rock red, brown or yellow by iron oxides set free by the alteration of basic silicates, or magnetite, and in the alteration of amphibolitic rocks to serpentine.

Where atmospheric waters percolate to depths considerably below the surface, their oxidizing power is diminished, and hydration is the principal alteration; the minerals thus formed through rearrangement of original constituents, and to a less extent through introduced substances, are epidote, chlorite, serpentine, pyrite, zeolites and quartz.

Leaching.—Leaching, usually accompanied by kaolinization, is a process of solution exercised by surface waters by which the soluble minerals are removed and the resistant minerals left as residual products. The minerals resistant to leaching and which commonly make up the leached upper parts of ore-bodies are quartz, kaolin, limonite and oxide of manganese. The completeness with which minerals are removed from the leached zone depends upon their relative solubilities and also upon the presence or absence of effective precipitants in that zone, which latter may cause residual masses of oxidized ores to form in the leached zone at the expense of the enrichments below.

In many ore-bodies the presence of active oxidizing precipitants causes the metals to be precipitated immediately or shortly after undergoing solution, a process known as pre-cipitation *in situ;* where this process predominates, secondary enrichments are prevented or are of relatively slight importance.

Under certain conditions limonite is completely removed, and the leached zone, white in color, is made up of quartz, kaolin and unchanged sericite. Where oxidation and solution of the accessory minerals proceed more rapidly than the solution and migration of the valuable metal, as often happens in deposits containing gold, the reduction in mass may result in a residual concentration; this form of enrichment is not uncommon, and is very different from the secondary enrichment due solution and migration.

Kaolinization.—Kaolinization is typically the result of surface agencies, except, perhaps, where hydrothermal solutions were acid in character. This process consists in the decomposition, and solution of feldspars and other minerals with the formation of kaolin as the residual product. A thorough kaolinization destroys the original character of the rocks attacked, and diverse rocks upon kaolinization yield products so similar that their original nature is distinguishable with difficulty, if at all. This process of alteration appears to depend upon the acid solutions set free by the oxidation of pyritic sulphides.

The presence of any considerable amount of kaolin is likely to indicate that sulphides have existed in the kaolinized area and that they have been removed in solution, perhaps with the formation of enriched sulphides in depth; kaolinization, there-fore, often constitutes a valuable guide in a search for secondary ore-deposits.

Theoretically, kaolinization is attended by a shrinkage in volume of from 12 to 15 per cent. which is partly offset by in-creased porosity. Kaolinization readily attacks feldspars, of which the soda-lime varieties appear to be the most susceptible, and microcline the most resistant; sericite also is readily decom-posed, and dark silicates are bleached through the removal of iron.

The typical product of complete kaolinization is a white,

soft, earthy mass, in which no minerals may be distinguished by the unaided eye except kaolin, quartz and limonite, and in which the structure of the original rock is completely destroyed. A rock that exhibits the outlines of the feldspar phenocrysts cannot be considered completely kaolinized, and such incomplete alteration rarely overlies important bodies of enriched sulphides. Kaolinization rarely extends into the wall rocks of veins, except where these rocks themselves originally carried a sulphide mineralization.

Oxidation.—Oxidation changes sulphides, arsenides, antimonides, tellurides and allied compounds into oxides, native metals, and salts containing oxygen, which, according to conditions of solubility and local precipitants, either remain *in situ*, or migrate to be redeposited under favorable conditions lower down in the deposit.

Access by oxidizing solutions is the controlling factor in this process, which reaches its most complete development in deposits carrying abundant sulphides, which through solution and removal, become progressively more porous under the action of the solutions and so help to bring the process to completion. Sulphides that are locked up as small grains within massive gangue minerals are rarely altered by oxidation, unless intense post-mineral shattering affords access to the solutions.

Oxidation tends to obscure primary structural relationships, and also to segregate the newly formed minerals into larger masses, either of enriched oxides or enriched sulphides.

Oxidation, through the more ready circulation of solutions, is likely to proceed to greater depths along veins than through the mass of the enclosing rocks; the country rock, therefore, is likely to exhibit signs of oxidation in the vicinity of veins, which is a valuable guide in underground exploration.

The order in which the various sulphides are attacked by oxidation varies according to their relative abundance, and also according to structural conditions and the character of associated minerals: Weed states[1] that the general order of

[1] *Trans.*, A. I. M. E., XXX, p. 429.

attack by oxidation is directly as the relative affinities of the several metals for oxygen, and inversely as their affinities for sulphur. The order in which the common primary sulphides are attacked, therefore, is arsenopyrite, pyrite, chalcopyrite, zincblende and galena.

Chalcopyrite in certain occurrences is quite resistant to oxidation, and pyrite varies in this respect markedly in different localities, occasionally remaining quite fresh for many years in mine dumps that are completely permeated by sulphate solutions, while elsewhere it has so strong an affinity for oxygen as to become highly heated when wetted. Tellurides appear to be ready subjects of oxidation.[1] Among the secondary sulphides, chalcocite is readily dissolved by acid solutions, as is illustrated in the various leaching processes. Tetrahedrite is a complex mineral of variable composition, and is irregular in its susceptibility to oxidation and solution. Most silver minerals are readily attacked by sulphate solutions, but become fixed in the presence of chlorine or its compounds.

The results of oxidation are likely to differ widely according to the mineral associations present, and are to a large extent subject to the influence of the enclosing rocks; deposits in limestone are likely to contain carbonates as oxidized ores, while silicates or oxides are likely to predominate in acid rocks. During the oxidation of sulphide deposits in limestone, acid solutions reacting with this rock produce gypsum, a relatively soluble but strongly combined compound that resists change or precipitation; it is probable that the soluble products of oxidation that are not precipitated leave but little trace for present observation. Contact minerals are quite resistant to the attack of oxidizing solutions, and frequently enclose unaltered sulphides at the surface, while associated bodies of originally more compact sulphides are oxidized to great depths. The oxidation of sulphide deposits in limestone is likely to be quite irregular, and to effect far-reaching rearrangements of minerals, as this rock is soluble and permits surface waters to form replacement

---

[1] Waldemar Lindgren, *P. P.* 54, U. S. G. S., p. 200.

deposits to an extent not approached in igneous rocks or quartzite.

A zone of rich oxidized ore is not uncommon immediately above the zone of sulphide enrichment, having been formed through the oxidation of the upper part of that zone. Not infrequently, and especially in desert regions, a deep zone is formed of partly oxidized, partly enriched sulphide, ore, which in most instances is not underlain by any definite zone of sulphide enrichment. In ores of the base metals an association of oxides with sulphides must usually be of smelting grade, as their widely differing specific gravities render concentration difficult.

The minerals that result from the oxidation and hydration of primary minerals are:[1] Chalcedony, opal, kaolin, limonite, pyrolusite, hematite, cuprite, the sulphates, carbonates, phosphates, silicates, and arsenates of the heavy metals, chlorides, and native gold, silver and copper.

The Migration and Enrichment of Metals.—During the oxidation of a deposit where one or more metals go into solution under conditions that permit migration, the metals commonly descend to the water level, where they are precipitated as secondary sulphides. This process of enrichment is probably most thorough and accomplishes its most important results in deposits whose primary ore consists in part of pyritic sulphides, which upon oxidation yield solutions of sulphates and free sulphuric acid to continue the solution and rearrangement. In deposits that contain important quantities of heavy sulphides the lower limit of the zone of enrichment is commonly well-defined.

The completeness of secondary concentrations of metals is usually proportional to the relative solubilities of their sulphides. Galena is relatively insoluble, and commonly does not migrate far, while copper, whose sulphate is readily soluble, frequently migrates through important vertical distances, and is commonly concentrated in well-defined zones.

The actual depths of secondary enrichments, and their

[1] Waldemar Lindgren, *Economic Geology*, II, p. 124.

relative depths in different parts of the same deposit, are determined by the position and character of the circulation channels and by the water level. Along the main channels of underground water circulation oxidizing and enriching solutions often penetrate to considerable depths below the ground-water level.

The waters that accomplish solution and enrichment of metals are commonly acid in character, and where observed, owe their activity to contained sulphuric acid and sulphates of iron. There is no lack of data, however, to show that most of the common sulphides are soluble in pure water, whose solvent action is greatly increased by small amounts of dissolved salts or acids, prominently among which may be mentioned carbonic acid, which in solution attacks pyritic sulphides.

Discussions of the reactions of solution and precipitation are of uncertain practical value, on account of the complex and constantly changing conditions under which they take place.

The character of the natural waters that flow or seep through the rocks is frequently a criterion upon which to base an opinion in regard to the probability of secondary enrichments in depth. Where the mine water is acid, or where it carries dissolved metals, the conditions are known to be those under which migration and enrichment take place: where the mine waters contain no acid or dissolved metals, and where there is evidence of precipitation *in situ*, there is but little likelihood of the existence of secondary enrichments in depth. Where water seeps through the rocks and appears in the workings, samples may be gathered for testing; where the rate of evaporation is greater than the seepage, efflorescences form on the walls of workings that may likewise be tested. Effloresences of the sulphates of iron, copper, zinc, magnesium, sodium and aluminum are of common occurrence on the walls of mine workings, especially in arid climates.

Pyrite and other primary sulphides are probably the most active precipitants of secondary minerals. The theoretical order in which metals are precipitated by sulphides from sulphate solutions is directly as their relative affinities for sulphur. The relative affinities of the common metals for sulphur is, according to Schuermann, mercury, silver, copper, bismuth, cadmium, antimony, tin, lead, zinc, nickel, cobalt, iron, arsenic and manganese. In a secondary rearrangement of a primary ore containing silver, copper, lead and zinc therefore the secondary sulphides should be arranged in the following order: uppermost, argentite, then chalcocite, bornite, chalcopyrite, galena, zincblende, and at the bottom pyrite.[1] The relative solubilities of the sulphates of these metals usually interfere with such an arrangement; lead sulphate, for example, is relatively insoluble, and commonly does not migrate far before being precipitated, while the more soluble metals descend farther before being precipitated. Organic matter is an active precipitant of metallic salts, and may be important as such in some deposits: precipitation by kaolin through adsorption appears also to be an important process in the formation of secondary ores.

Upward Migration.—The upward migration of solutions due to capillary action is occasionally a factor in the rearrangement of values in the upper parts of ore-deposits. Limonite is sometimes formed at the surface through the oxidation and evaporation of solutions carrying sulphate of iron, and occurrences are known where the chloride of silver owes its distribution to upward migration.

AT LEADVILLE, COLORADO.[2]—At a point where an ore-body had been to a large extent removed by erosion the loose wash for a distance of 10 ft. above the solid rock was found to carry chloride of silver; several hundred tons of this material were shipped that carried from 40 to 60 oz. of silver. This deposit was found at a depth of 150 ft. below the surface.

The Influence of Circulation Channels on Secondary Alterations.—Secondary alterations are in every case controlled by the channels that permit surface waters to circulate through the rocks and ores. In massive ores in districts where heavy rainfall and steep slopes give rise to an erosion more rapid than percolation and alteration, these processes are of no effect, and

---

[1] Waldemar Lindgren, *P. P.* 43, U. S. G. S., p. 182.
[2] Max Boehmer, *Economic Geology*, III, p. 340.

primary ores and minerals appear at the surface. Where post-mineral fracturing permits surface waters to pass through the ores, however, a thorough rearrangement of the minerals and values may be accomplished, even in districts of heavy rainfall, steep slopes, and rapid erosion. Solution is a characteristic feature of the action of surface waters upon the upper parts of ore-bodies, which commonly become more and more porous under their attack and through seepage a supply of water is thus maintained under hydrostatic pressure for the continuance of the process, even in districts of scanty rainfall. It is thus seen that conditions of heavy or slight rainfall, with attendant rapid or slow erosion, may be balanced or counteracted by the presence of good or poor circulation channels, and enrichments may form under either set of conditions. Under conditions of heavy rainfall, sulphide enrichments are likely to form without important zones of oxidized ores; under desert conditions, the oxidized zone is likely to be deep and important though characteristically irregular, often merging irregularly into the sulphide enrichment. The subject of circulation channels will be further discussed in the chapter on secondary ores and ore-shoots.

The Relation of Oxidation and Enrichment to Topography and Water Level.—A close relation exists between the depth of secondary alteration and the ground water level. Oxidation commonly reaches to or somewhat below the water level, and enriched sulphides commonly occur at or just below this horizon. While not a uniform relation, the water level commonly follows the topography.

The ground water may be considered to fill a complicated but connected system of porous rock masses, fissures, joints, and other open spaces. In most instances the water level reflects the surface in a modified form; being deepest below the middle slopes, and becoming shallow again in the foothills, where the water reaches the surface in springs. In the Southwestern States, the valleys are commonly filled with débris; the apparent foothills, therefore, are often the middle slopes as referred to the outlet of ground water, and oxidation may proceed to important depths beneath them.   The ground-water level depends upon the permeability of the rocks traversed as well as upon the height of outlet.   The water level should be horizontal beneath a hill of loose conglomerate, for example, but might easily reach the surface in a tight rock that does not permit seepage.

In investigating the probable depth of the water level in a new district, it should be borne in mind that springs may exist at levels considerably above the general ground-water level, being the mouths of accidental drainage channels that permit a more easy escape of waters than does continued seepage.   A number of springs at approximately the same elevation is commonly a sign of the true water level.

Where there has been much faulting, a district may be divided into fault blocks, each of which constitutes a separate hydrostatic basin, the water being impounded to a ceratin extent by the impervious gouge along the fault planes.   The groundwater level is likely to be deep in arid regions, and shallow in regions of heavy rainfall, and the same relation holds for the depths to which oxidation may be expected to penetrate.

Where the water level is permanent, well-defined, and not too deep, oxidation and enrichment tend to form relatively regular zones; where the water level is irregular or practically lacking, oxidation and enrichment are likely to be quite irregular.   The water level in any district is likely to change with time, and the level at some past epoch may have been the determining factor in the location of enrichments as now found.

The Irregularity of Oxidation and Enrichment.—A large mass of primary ore uniformly mineralized and brecciated, or of uniform porosity, should yield upon oxidation and enrichment bodies of secondary ores of uniform depth that bear a like definite relation to the topography.   This result, rarely attained, finds its nearest expression in certain disseminated chalcocite enrichments.

Oxidation and enrichment persist deepest along the more prominent zones or lines of fracture, and it is not unusual to find tongues of secondary ores deep within the primary zone.   Furthermore, homogeneous masses of unaltered primary sulphides are often found in the oxidized zone, having been protected by their massive structure against attack by oxidizing and dissolving solutions. In a search for secondary ores, therefore, the post-mineral fractures should be carefully studied, and exploration should be directed to expose their intersections with the zones of primary mineralization.

It occasionally happens that the rôle played by fracturing in the formation of secondary enrichments is emphasized by the occurrence of such enrichments along post-mineral fissures or joint planes, the precipitation having been accomplished by absorption, or other agencies.

The Zones Developed by Surface Agencies.—The ideal result of the rearrangement of a primary ore-body by surface waters would be the formation of a surface zone of resistant minerals, leached of their metals, beneath which would be found an oxidized zone of low grade, and in turn at successively greater depths, a zone of rich carbonate and oxide ores, and a zone of secondary sulphide enrichment resting upon the primary ores. The complete series is occasionally developed, but most deposits exhibit irregularities due to structural features, relative rapidity of erosion as compared to solution and concentration, differing solubilities of the minerals present, and the precipitative action of associated rocks or minerals.

The oxidized and leached zones are usually deep in arid regions, and an important proportion of the valuable ore is likely to occur in the oxidized zone; in wet climates, the leached and oxidized zones are commonly slightly developed, and the major proportion of the valuable ore is likely to be found as sulphide enrichments.

In most cases the enriched ores are not the result of the leaching of the present oxidized zone, but are derived from a great vertical extent of vein or deposit, the residual minerals of which have been removed by erosion; the present enrichments may, therefore, be considered to represent repeated reconcentrations.

The actual depths of the various zones vary widely in different districts, and in different parts of the same district, or even in the same deposit, depending upon the local conditions of water circulation. The depths at which changes commonly occur in other districts do not form reliable guides in the examination

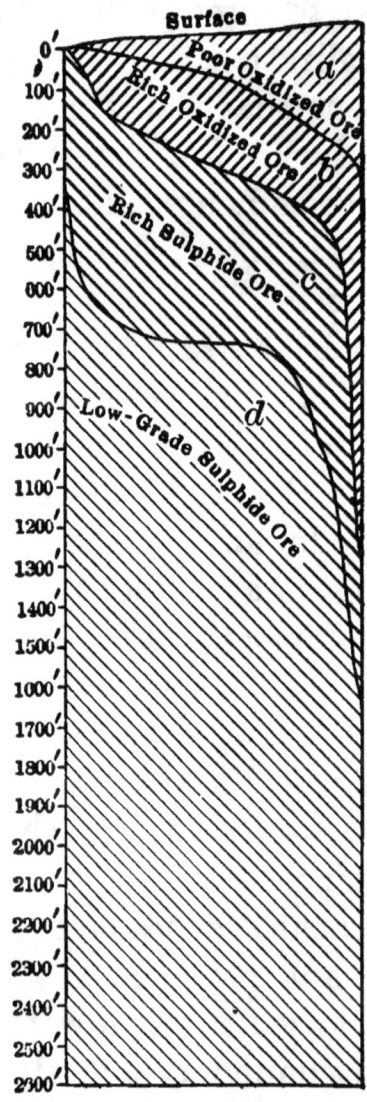

FIG. 71.—Diagram showing the occurrence of oxidized, enriched, and primary ore in the Granite vein, Phillipsburg, Montana; the horizontal dimensions of the areas show approximately the relative amounts of each type of ore at various depths, but are without significance as to their positions in the vein. *After W. H. Emmons.*

of a new deposit. The several zones occasionally are sharply marked, and the change comes without warning, but commonly the zones grade gradually one into the next, and so give warning of the impending change.

While it is not possible to predict the depths at which the several zones will occur, it is possible and of great practical importance to recognize the zone exposed, and so to appreciate the changes that may reasonably be expected with deeper exploration. The relative importance of the several zones depends in large measure upon the minerals and metals of the primary ore. Copper, for example, is frequently completely leached from the surface zone and is found as chalcocite enrichments without overlying residual ores. Where oxidation proceeds more rapidly than solution and enrichment, as for example, in gold deposits, the oxidized ore commonly contains the bulk of the values of the primary ore, perhaps in concentrated form, while the zone of enriched sulphides is poorly developed, or lacking. The upper part of the enriched zone is commonly the richest horizon of a secondary deposit.

In general, the completeness with which any ore-body is rearranged by oxidation is proportional to the relative solubilities of the metals it contains in the presence of existing precipitants, and the degree of post-mineral fracturing, or of porosity due to abundant sulphides as compared with resistant gangue minerals.

In the upper zones of deposits that have been subject to oxidation and enrichment, kernels, or residual masses, are likely to be found that represent the character of the zones below. Residual nodules of oxidized ores are likely to occur in the leached surface zone, nodules of enriched sulphides in the oxidized zone, and masses of residual primary ore in the zone of sulphide enrichment. A study of such residual masses may yield information in regard to the character of the lower zones not yet exposed by exploration; a free milling oxidized gold ore, for example, may contain masses of unaltered and refractory pyritic ore, indicating that such will be the nature of the primary ore when reached. Residual kernels of galena frequently carry higher values in silver than does this mineral in the primary ore.

At Phillipsburg, Montana, the Granite Vein[1] affords a good

[1] W. H. Emmons, *Bull.* 315. U S. G. S., p. 42.

example of the rearrangement of values by surface agencies. The vein has suffered post-mineral movement and much of the ore is fractured, permitting ready access to percolating surface waters. The primary ore consists of pyrite, arsenopyrite, tetrahedrite, and tennantite, with some galena and zincblende. Sparingly scattered through this ore are small specks of pyaragyrite, proustite, and lesser amounts of realgar and orpiment. The primary ore carries from 20 to 30 oz. of silver and from $1.50 to $3.00 in gold; the gangue is quartz and rhodochrosite. The secondary or enriched sulphide ore, resting upon the primary ore, consists of bands of quartz and rhodochrosite alternating with bands of rich sulphides, comprising argentite, proustite, pyrargyrite, tetrahedrite, tennantite, pyrite and arsenopyrite. Galena and zincblende are locally abundant; chalcocite and bornite are of rare occurrence. This ore carries from 50 to 1000 oz. of silver and from $4 to $8 in gold. Probably more than one-half of the silver in this ore is present as dark ruby silver, or pyrargyrite, with some proustite, occurring in minute veinlets or seams filling cracks in the vein, as films on the outside of crushed vein material, or as relatively large crystals lining cavities. This ore is responsible for a large proportion of the dividends that the mine has paid. The enriched oxidized ore occurs just above the enriched sulphide ore; it is composed of quartz stained by the oxides of iron and manganese, and less commonly by copper carbonates. Cerargyrite and native silver occur as thin seams cutting through the quartz and as films on the outside of quartz fragments. Associated with this ore are small quantities of argentite and pyromorphite, and residual galena, zincblende, pyrite and chalcopyrite are of local occurrence. This ore carries from 300 to 400 oz. of silver and from $5 to $16 in gold. The poor, or leached, oxidized ore forms the upper zone; it is composed of quartz, commonly broken and stained by iron and manganese oxides. It carries small quantities of lead carbonate, malachite, azurite, chrysocolla, pyromorphite, and residual pyrite and galena; it carries less than 40 oz. of silver, and but little gold. The sulphides contained by the oxidized ores are remnants that have escaped oxidation, their massive structure having prevented the access of the oxidizing solutions; low-grade pyrite occurs locally as druses of secondary origin.

The Depth of Vein Leached to Form Existing Enrichments.— The metallic content of the enriched ores in any deposit is rarely due to the leaching of the present depth of the oxidized and leached zones, but is probably derived from a large vertical extent of vein now removed by erosion. It is occasionally possible to calculate approximately the depth worked over by surface agencies necessary to have produced the existing enrichments.

At Phillipsburg, Montana, the Granite Vein[1] yields data upon which to calculate the depth of primary ore worked over necessary to account for the enrichments found. Here the primary ore carries about 25 oz. of silver. The enriched ores, about 400 ft. in vertical extent in one part of the mine, average 175 oz. of silver. On the basis of these figures, assuming a constant width, the depth of primary ore necessary to produce the enrichment, must have been about 2400 ft., provided that all of the contained silver found its way into the enriched ores.

[1] W. H. Emmons, *Bull.* 315. U. S. G. S., p. 43.

CHAPTER IX

RESIDUAL ORES AND THEIR DISTRIBUTION

The Precipitation of Ores in The Zone of Oxidation.—In deposits in which the secondary zones are well-defined a layer of rich oxidized ore is frequently found to immediately overlie the enriched sulphides, from which it is derived by direct oxidation in place. In deposits that contain oxidizing percipitants particles and masses of residual oxidized ore are likely to be scattered through the leached zone, having been precipitated during migration before reaching the zone of enrichment. This precipitation hinders secondary enrichment and in extreme cases prevents the formation of such concentrations.

Residual ores are precipitated by various reagents. Carbonates are formed through the action of the carbonic acid contained in surface waters, and also directly through the replacement of calcite, especially where the containing rock is limestone. Native metals, such as copper, gold, and silver, are formed by the action of reducing agents, among which organic matter and ferrous sulphate are prominent. Kaolin, gouge, and certain shales occasionally act as powerful precipitants through adsorption. Silver is often found as residual chloride, formed through precipitation by the chlorine contained in surface waters.

Residual ores are often the result of incomplete solution, the relatively insoluble minerals being left behind during the migration of associated metals; incomplete oxidation and solution often leave residual masses of unaltered, or partly altered, sulphides in the oxidized zone. Such residual particles or masses have commonly been enriched by additions from circulating, metal-bearing solutions.

During oxidation under conditions that permit reprecipitation, such as the oxidation of sulphide deposits in limestone, a scattered primary mineralization is probably often segregated into masses of rich oxidized minerals without important vertical migration. In the zinc and lead deposits of the Mississippi Valley the lead is usually found as residual enrichments of galena, above the water level, while the zinc and iron sulphides have in great part been removed by solution and redeposited at and below the water level.

Oxidized Ores of Copper.—Native copper is the last stage in the reduction of copper compounds, and is frequently found as pellets or films in the upper oxidized zones of copper deposits. It is commonly associated with cuprite, from which it is probably derived in most cases. Cuprite is often found just above the chalcocite zone, where it is formed from the chalcocite by direct oxidation. Both native copper and cuprite are commonly indicative of long and thorough oxidation, and are frequently found above important chalcocite enrichments. Tenorite, or melaconite, is intimately related to and associated with chalcocite. Chrysocolla is a common residual ore of copper, and is more likely to be abundant in siliceous rocks than in limestone, where carbonates are likely to prevail. The silicate of copper is frequently associated with manganese in black compounds of indefinite composition commonly called copper-pitch ores. Malachite and azurite are among the most important oxidized copper minerals; they are relatively resistant to oxidizing processes and are frequently found as residual ores. Azurite appears to be the more resistant of the two. Malachite and azurite frequently occur with limonite, an association that is explained on the supposition that siderite was formed with the copper carbonates, but subsequently, being more susceptible to alteration, decomposed to limonite. Brochantite is occasionally an important oxidized ore of copper, being often a transitional step in the formation of carbonates. In desert climates, atacamite may constitute an important ore, though its easy solubility confines it to regions of extreme aridity.

Oxidized Ores of Lead.—The only important primary ore of lead is galena, which is relatively resistant to oxidation, and is frequently found near the surface. Upon oxidation it alters slowly to sulphate and carbonate, relatively insoluble products that commonly do not migrate far before being precipitated. Under the attack of sulphate solutions galena alters to anglesite, which may remain as a residual ore; more frequently the anglesite is altered to cerussite, which is the less soluble compound. In certain districts, as for example, the Coeur d'Alenes[1] where galena is in primary association with siderite, the latter mineral upon alteration to limonite sets free abundant carbonic anhydride, and the galena passes into cerussite apparently without the formation of anglesite.

Oxidation occasionally proceeds to great depths in lead deposits, and the bulk of the ore is found as residual particles or masses that have been added to somewhat by enrichment through solution and precipitation. Pyromorphite, and the oxides of lead, plattnerite and massicot, are of less common occurrence, and are usually characteristic of the upper parts of the oxidized zone, in which lead chromate and molybdate also are occasionally found.

Where residual masses of galena occur in the oxidized zone they frequently carry silver due to enrichment by migrating solutions, and are higher in grade than the primary galena unaffected by surface processes.

Oxidized Ores of Zinc.—Sphalerite, or zincblende, the only important primary zinc mineral, yields readily to oxidation with the formation of zinc sulphate, a very soluble salt that commonly migrates far before being precipitated.

It is rare to find any traces of zinc in the upper parts of oxidized ore bodies; lower down in the deposits, zinc is frequently precipitated as calamine or smithsonite replacing limestone or other sedimentary rocks. These minerals are difficult to identify, as they usually reproduce with great fidelity the structure of the replaced rock, and their low specific gravities fail to call attention to their high metallic content.

[1] F. L. Ransome, *P. P.* 62, U. S. G. S., p. 132.

At the Horn Silver Mine, Utah,[1] a secondary distribution of metals has taken place under conditions of partial oxidation. The upper 400 ft. of the ore-body, whose gangue is chiefly quartz and barite, carries lead-silver ores, the lead as galena, cerussite, anglesite and oxides, and the silver as cerargyrite and as ruby silver; zinc and copper are scanty or lacking. At 700 ft., large bodies of zinc ores occur as carbonate and silicate associated with some lead. From 650 to 750 ft. copper comes in, occurring as chalcocite, with which is associated galena.

Residual Shoots of Gold Ores.—The upper parts of gold deposits are commonly richer than the underlying primary ores. This residual enrichment in the zone of oxidation is in large degree owing to the solution and migration of associated minerals, the reduction in mass giving rise to a concentration of the undissolved gold. A further factor in this concentration is the actual solution, migration and enrichment of gold.

The solution and migration of gold takes place more slowly in most cases than the migration of associated metals; this results in the zone of gold enrichment being left at a horizon above the enrichments of associated metals. The minerals that commonly occur with gold in the upper parts of oxidized ore-bodies are quartz and limonite; the limonite is often removed by solution, leaving the gold in a porous mass of quartz, the result being a still further concentration of the gold with respect to the containing gangue.

A surface enrichment frequently takes place in gold deposits during the disintegration of the outcrop on weathering; the lighter grains of gangue are washed or blown away and the gold particles tend to work downward into superficial cracks, forming enrichments that usually extend for very short distances only below the surface. Where erosion or glaciation proceeds more rapidly than oxidation, the soft upper parts of ore-deposits are removed and primary ores outcrop at the surface.

Shoots of residual gold ores must be considered as surface phenomena. Their greatest dimension is more likely to be

[1] S. F. Emmons, *Trans.*, A. I. M. E., XXXI, p. 658.

horizontal than vertical, being confined to the zone above the water level, or the depth to which oxidation has penetrated.

The Distribution of Silver by Oxidation.—Silver ores are affected in various ways by oxidation. The sulphate of silver, being readily soluble, migrates freely, and may precipitate as sulphide enrichments at the water level. Precipitation of silver as chloride is frequent, however, in the zone of oxidation, especially in arid regions where the residual ores are likely to be the most important: finally, silver chloride is slightly soluble, and, migrating through relatively short distances, is likely to produce local enrichments in the oxidized zone. A residual concentration of silver as chloride and native silver may also take place through the removal by solution of relatively soluble gangue minerals. Minerals that are frequently associated with cerargyrite in oxidized ores are native silver, the bromide and the iodide.

The residual ores of silver, being confined to the zone of oxidation, commonly show a distribution related to the topography.

The Distribution of Manganese by Oxidation.—The oxides of manganese are frequently found as residual concentrations in the oxidized zone. Indeed, it is sometimes difficult to explain their abundance in the oxidized parts of certain deposits whose primary ore contains but little of this element. The oxides of iron and of manganese frequently occur with residual concentrations of both silver and of gold. Manganese oxide usually is most abundant in the upper part of the oxidized zone.

The relative abundance of manganese in the oxidized zone as compared with the primary zone often gives an idea in regard to depth of leaching and erosion necessary to produce such concentrations.

CHAPTER X

SECONDARY ORES AND ORE-SHOOTS

The Criteria of Secondary Ores.—Secondary minerals are the result of a process of concentration and enrichment and are commonly richer than the primary minerals of the same deposit. Secondary ores that contain abundant sulphides are commonly distinguishable in the field from primary ores; this distinction is more difficult with ores that carry a small quantity only of sulphide minerals, and microscopical examination of the ore in thin section is advisable in all doubtful cases.

It is frequently difficult, and in some cases impossible, to determine the primary or secondary character of quartzose gold ores. The safest guide is the presence or absence of gaseous or fluid inclusions in association with the ore minerals; such primary association coupled with a lack of traces of oxidation is indicative of primary origin.

Any ore that carries traces of oxidation, such as iron stains, dendritic oxides of manganese, kaolin and so forth must be considered to have been acted upon by surface agencies, and secondary enrichment must be suspected. Further examination, however, may indicate that the contained values are primary, and, therefore, persistent in their vertical distribution.

The manner of occurrence of secondary minerals is frequently characteristic; secondary minerals, being due to the action of surface waters, are commonly found in cracks through earlier minerals, or as crusts lining vugs or surrounding nucleal masses, or in general, in some distribution later than and not conforming to the arrangement of the primary minerals. Secondary sulphides are often found as soft black pulverulent masses or coatings, although they frequently occur well-crystallized or as compact masses of metallic luster. Where the relationship

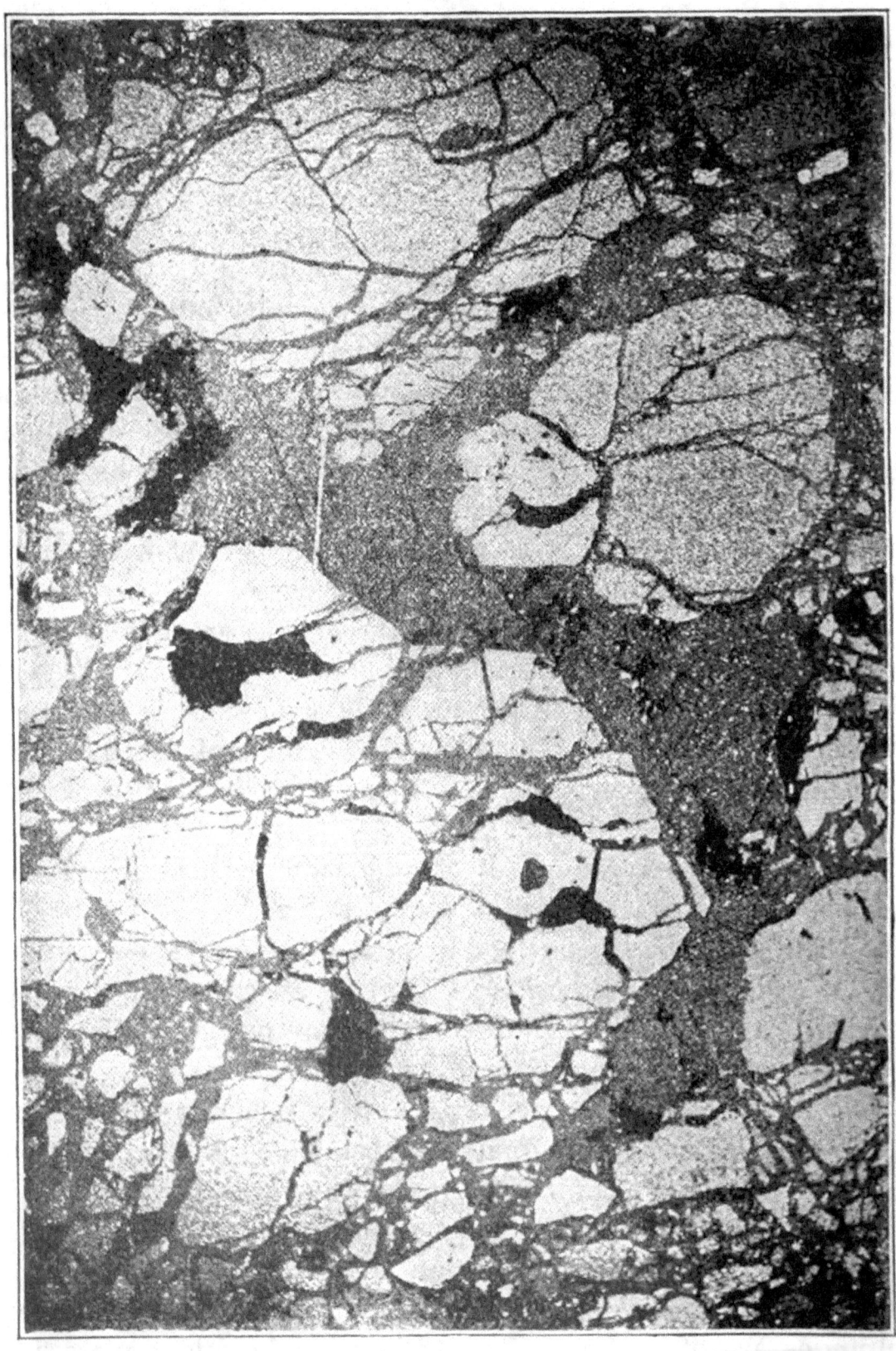

FIG. 72.—Photomicrograph of secondarily enriched ore from the lower part of the chalcocite zone, Morenci, Arizona, showing the occurrence of the later and richer sulphide through the pyrite. Dark gray = secondary chalcocite, developing by replacement in pyrite; light gray = pyrite; black

may be determined, secondary ores are always found in connection with post-mineral fractures, or beneath a porous capping.

The form of an ore-deposit often indicates its secondary origin. This criterion, however, is not available until after the deposit has been partly explored. An ore-deposit formed above and resting on an impervious stratum is probably the result of descending waters and is of secondary origin. A deposit that

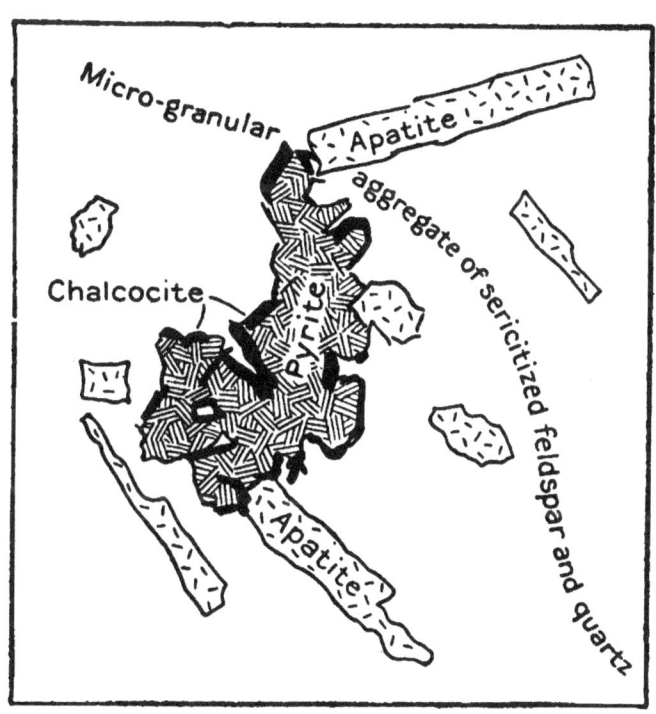

FIG. 73.—Group of pyrite crystals, showing secondary chalcocite along the edges. *After Paige.*

branches as it is followed downward, leaving the channel of primary mineralization to become enclosed in the wall rocks, is certainly of secondary origin.

Most veins are lines of weakness likely to be reopened by post-mineral fractures; surface waters frequently find their way to great depths along such fractures and secondary processes may be active to any depth to which they penetrate. The range of influence of these processes may be appreciated upon considering the number of deep mines that are wet.

A list of the minerals that are formed by oxidizing processes is given in a preceding chapter. Many minerals that are found as secondary enrichments are also formed by primary processes, and their occurrence, therefore, may not be accepted as proof of secondary origin. Minerals that are secondary in a great majority of their known occurrences are covellite, chalcocite, kaolin, chalcedony, and the sulpharsenites and sulphantimonites

MINERALS FORMED IN LOWER GROUND-WATER ZONE

(Zone of Secondary Sulphide Enrichment[1])

Quartz	Bornite
Chalcedony	Covellite
Opal	Chalcocite
Kaolin	Argentite
Gold	Pyrargyrite
Silver	Stephanite
Pyrite	Polybasite
Galena	Other complex sulpharsenites
Chalcopyrite	and sulphantimonites.

Secondary Ore-Shoots.—Secondary ore-shoots are the results of surface agencies and bear a definite relation to the surface, and in any district they are likely to be found within certain limits of depth. In shallow deposits, the vertical distribution frequently reflects the topography, while deep enrichments commonly occur in more nearly horizontal zones. The greatest dimension of secondary ore-shoots is, in general, horizontal, while in primary ore-shoots the reverse is the rule. Furthermore, in secondary deposits the change in values is likely to be most pronounced vertically, while in primary deposits the best defined changes in value are commonly in the direction of their strike.

Secondary processes affect in like manner all types of primary deposits, the degree of alteration and reconcentration depending upon the original constituents of the deposits and upon the structural features that control the passage of waters. In the

[1] Waldemar Lindgren, *Economic Geology*, II, p. 124.

consideration of secondary ore-shoots the primary distribution of values must not be lost sight of; secondary enrichments commonly occur at certain horizons within the primary ore-shoots, although cases are not rare where secondary enrichment has affected veins for long distances along their strike, the primary mineralization of which was probably irregular.

In exploration for flat-lying or bed-like enrichments, such as the common type of disseminated chalcocite enrichments, the workings should be vertical, so as to pass through the successive zones. In exploration for possible tongues of secondary ores

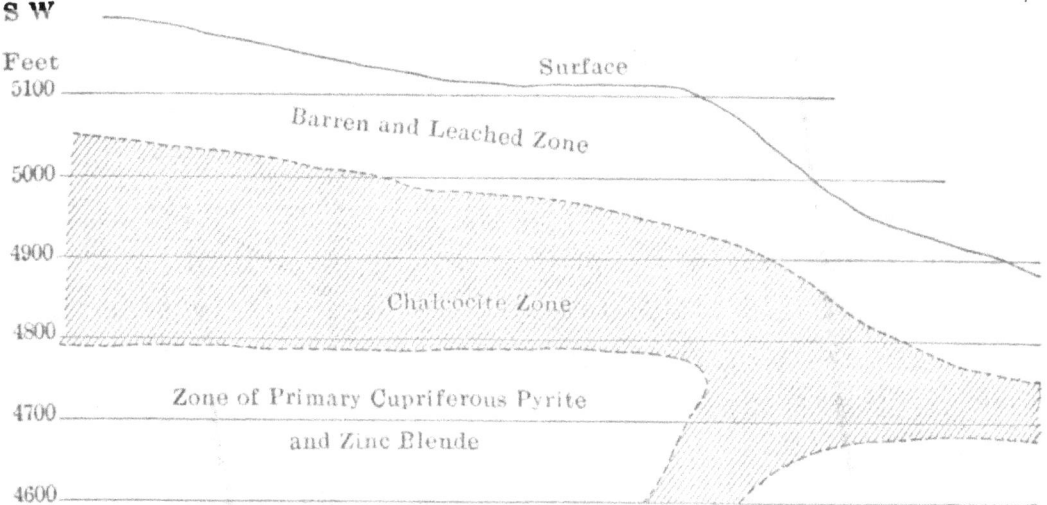

FIG. 74.—Longitudinal section of a part of the Humboldt lode, Morenci, Arizona, showing the relations of the leached, enriched, and primary zones. *After Lindgren.*

along post-mineral fractures in the zone of primary mineralization, however, the workings should be horizontal; vertical exploration within the zone of primary ores is futile.

In the investigation of a secondary ore-deposit it is wise to average all samples taken from each level; a comparison made between the different levels will often indicate the trend of increase or decrease in values as depth is attained, and so perhaps afford a basis upon which to form an opinion of the continuity of the ore-shoots in depth.

AT MORENCI AND METCALF, ARIZONA, the important chalcocite enrichments all occur high up on the hills; erosion appears to have proceeded faster than leaching and enrichment, and the

canyon bottoms are all in rock carrying a primary pyritic mineralization only.   The elevation of the surface is here a guide in
exploration; the higher slopes are considered promising ground,
while explorations from the canyon bottoms cannot be expected
to yield results.

The Effect of Structural Features on Secondary Ore-Shoots.—
Post-mineral fractures are the best guides in a search for secondary ore-deposits; these deposits are in every case due to surface
waters, which circulate most abundantly along fractures, faults,
brecciated zones, joint planes, or other openings.   Any search
for secondary ore-bodies, therefore, should be directed to explore
the intersections of post-mineral fractures with the general
trend of the primary mineralization.   Solid, impervious masses
of primary ore are not subject to enrichment, and commonly carry
their primary mineralization unaltered close to the surface.
Thorough reopening of fractures during oxidation constitutes
the most favorable condition for the formation of secondary
enrichments.   The general association of secondary ores with
fractured, shattered, or permeable rocks cannot be too strongly
emphasized.

The Effect of the Water Level on Secondary Ore-Shoots.—In
evenly shattered or uniformly permeable deposits precipitation
and enrichment generally take place at the water level, where
chemical conditions change and precipitants in the form of
primary sulphides are likely to be met.   Where, however, a
prominent fracture cuts a primary ore-deposit, and forms an
important channel for descending solutions, secondary enrichments are likely to persist deep into the zone of primary ores.
The most favorable conditions for large secondary enrichments
are a repeated reopening of the fractures coupled with a gradual
sinking of the water level.

**The Effect of Chemically Active Wall Rocks on Secondary Ore-
Shoots.**—In deposits that contain active precipitants, or that
are contained within rocks having like effect, migration and
enrichment do not take place, the metals being precipitated
immediately upon going into solution.   Where a deposit traverses

different rocks one of them may permit migration and enrichment, while the others may impede or stop these processes.

Acid igneous rocks and acid gangue minerals appear to be favorable to migration and enrichment, while sedimentary rocks and carbonate gangue minerals appear to be unfavorable. Contact deposits, on account of their resistant minerals and their commonly strong precipitative action, rarely contain secondary enrichments, although such enrichments are not rare in associated intrusives.

At CLIFTON, ARIZONA.[1]—The veins where contained in porphyry carry important secondary chalcocite enrichments, but where they enter limestone or shale, they become impoverished.

The Effect of Porosity on Secondary Ore-Shoots.—Most fresh rocks are massive and impervious, except certain sediments, whose active precipitative action prevents migration and enrichment. Alteration, usually involving kaolinization, is therefore a usual prerequisite of enrichment. The ease with which rocks or gangue minerals alter to a permeable mass is an important factor in secondary enrichment, aside from the existence of the necessary post-mineral fracturing that gives access to the surface waters.

The Effect of Primary Mineralization on Secondary Ore-Shoots.—Disseminated enrichments are commonly confined to areas whose outcrops bear traces of abundant primary mineralization. The tendency in vein-like deposits, however, is toward the confinement of descending and enriching waters within the vein, and thus produce relatively long enrichments whose distribution is not so definitely controlled by the primary ore-shoots.

Vein deposits are commonly more persistent vertically than brecciated zones, and enrichments in veins may be due to the leaching of many thousands of feet of the vein now removed by erosion, resulting perhaps in a great concentration, and a wide difference in value between the primary and the secondary ores. In disseminated deposits there is usually less difference in value between the primary and the secondary ores.

[1] Waldemar Lindgren, *P. P.* 43, U. S. G. S., p. 204.

Ores Containing Both Sulphide and Oxidized Minerals.—In arid regions where the water level is deep and the supply of surface water is irregular, a zone is frequently developed in which the valuable metals occur in both sulphide and oxidized form, with, in general, slight enrichment only of the primary values. The occurrence of such ores through a considerable vertical range indicates incomplete solution and enrichment, and in most cases, the absence of any well-defined zone of secondary sulphides at greater depth.

The Enrichment of Copper.—Primary copper ores commonly contain abundant pyrite, and are actively attacked by oxidation; the sulphates thus formed are very soluble, but yield their copper readily as secondary sulphides upon coming in contact with fresh pyrite at the water level. As a result of these conditions, copper is peculiarly susceptible to solution and enrichment, and the best defined types of secondary deposits are those of copper.

Chalcopyrite is the usual starting-point, being the most abundant primary copper mineral, and is the form in which the copper of cupriferous pyrite is supposed to exist. Bornite, enargite, tetrahedrite, tennantite and occasionally chalcocite are primary copper minerals of lesser importance. The secondary copper sulphides are chalcocite, chalcopyrite, enargite, bornite and covellite; the last named being probably exclusively secondary in origin.

On account of the strong precipitative action of calcite, chalcocite enrichments are more rare in limestone than in other rocks, except where the primary sulphides occur in large masses.

Chalcocite Enrichments.—The final stage in copper enrichment is chalcocite, which is also economically the most important copper mineral. Kaolin and quartz are the typical associates of secondary chalcocite, which commonly also encloses residual pyrite. The upper part of the chalcocite zone is commonly the richest.

The zones developed in massive pyritic deposits are usually well-defined, and the changes from oxidized ores to chalcocite,

and from chalcocite to pyrite are usually sudden; the enrichment is often a shallow zone of nearly pure chalcocite. Pyritic
masses that have been shattered commonly carry their secondary
chalcocite in seamlets, or as coatings on nucleal masses, and enrichment in this form may persist through considerable depths.

At Bisbee, Arizona,[1] the primary mineralization of lean
cupriferous pyrite occurs in large bodies in limestone; the country
rock is commonly impregnated with pyrite, and minute particles
of contact minerals in the vicinity of the ore-bodies, and also
in places exhibits a partial silicification. Secondary processes
have transformed and enriched these primary deposits, resulting
in the formation of ore-bodies of great economic importance. In
typical occurrences a core of lean cupriferous pyrite is surrounded
by a shell of pyrite carrying secondary chalcocite, commonly as
seamlets and coatings on the pyrite grains; this shell is surrounded
in turn by an envelope of ferruginous clays containing oxidized
copper ores. Where prominent fractures cut the pyritic masses
they are accompanied by chalcocite, and where several such
fractures occur close together large and rich ore-bodies result. In
exploration, when a drift through one of these pyritic masses
shows a gradually rising copper content, it is known that the
periphery of the deposit is being approached.

At Ducktown, Tennessee,[2] the outcrops consist of hydrous
iron oxide associated with kaolin and quartz. This material,
which is a valuable iron ore, extends to a maximum depth of
about 100 ft. Below this iron ore there is commonly a few
feet of chalcocite ore, which in most of the deposits lies like a
floor between the gossan and the underlying primary sulphides;
the primary ore consists of pyrrhotite, pyrite and chalcopyrite
associated with lesser amounts of zincblende, bornite, specularite
and magnetite, with a gangue of lime silicates, quartz and
marmorized limestone. The enclosing country rock is schist.

At Clifton, Arizona, the Copper King vein exhibits well the
several zones produced by surface agencies. The outcrop of

[1] F. L. Ransome, *P. P.* 21, U. S. G. S., p. 155.
[2] W. H. Emmons and F. B. Laney, *Bull.* 470, U. S. G. S., p. 169.

this vein is white quartz, showing in a few places casts of originally contained sulphides, but for the greater part, the quartz is massive, and unstained by iron. This outcrop carried, where underlain by the ore-body, a little malachite and chrysocolla as stains on post-mineral fractures through the quartz. Beneath the hard capping, kaolin is mixed with the quartz, which carries copper stains and small patches of residual carbonates and silicate of copper, which gradually increase in quantity until the vein becomes ore. Beneath this oxidized ore, chalcocite, associated with kaolin, quartz, and cuprite, comes in, gradually changing to chalcocite, associated with pyrite and quartz, which form a very hard and tough ore for several hundred feet in depth. The pyrite in this ore becomes more prominent as depth is attained, until upon the disappearance of the chalcocite, the chief values in the pyritic material are as chalcopyrite, which in turn becomes less in depth, and in the lower part of the deposit the primary ore is exposed, consisting of pyrite, associated with a little chalcopyrite and zincblende in a gangue of quartz and quartz-porphyry. The pyritic ores carry low values in gold.

Disseminated Chalcocite Enrichments.—Disseminated chalcocite enrichments, popularly called "porphyry copper deposits," now form one of the principal reserves of known copper ores; many millions of tons of such ores averaging perhaps 2 per cent. copper have been developed at different camps in the Southwestern States.

The primary ore of this type of deposit is disseminated cupriferous pyrite, usually associated with quartz and accompanied by sericitization of the containing rock. This mineralization is commonly introduced into crushed or sheared zones, the minerals forming veinlets through the rocks; it also frequently follows joint planes, and occurs as scattered particles through the rock between such fractures.

Upon the oxidation of the pyrite, solutions containing the sulphates of iron and of copper and sulphuric acid seep downward through the mineralized mass, kaolinizing the rock and leaching it of its contained copper, which is precipitated on coming in contact with unaltered pyrite at the water level. This process, while very irregular over short distances, produces enriched deposits that are relatively uniform when considered in large masses. The enrichments commonly bear a definite relation to the depth below the surface and to the ground-water level.

In some instances a part of the iron is left behind in the leached zone as limonite, while in others the iron is completely removed with the copper, the leached capping being a porous mass of white kaolin and quartz. In the upper oxidized zone, where not completely leached, large bodies of low-grade residual ores are frequently found, in which the copper occurs as chrysocolla, carbonate, or partly oxidized particles of sulphide minerals, and also in some cases, as native copper and cuprite. The low grade of these residual ores coupled with the presence of perhaps an important proportion of the contained copper as sulphate results in serious metallurgical difficulties, due to solution and to the low specific gravity of some of the copper compounds.

Secondary chalcocite enrichments, while varying greatly in their geological associations, present a remarkable similarity to casual observation, the secondary processes having tended to produce like results from dissimilar primary deposits. Deposits of disseminated copper minerals are in many cases associated with monzonitic intrusives.

Important features in the consideration of these deposits are the depth at which they occur, whether susceptible to stripping and steam-shovelling, and the strength of their walls, which are commonly kaolinized, soft, and likely to cave upon the removal of any large quantity of ore.

Inasmuch as these deposits are commonly of horizontal tabular form, exploration should be carried out vertically. The zone of enrichment is expected beneath a thoroughly leached and kaolinized capping, and is underlain by primary pyritic ores, into which it is futile further to continue vertical exploration.

The primary ore may usually be recognized by the freshness of the containing rock, as well as by the typical primary minerals. In certain cases disseminated enrichments have been found beneath pyritic material, but always where this pyritic material was disintegrated and the associated or containing rock well kaolinized, the pyrite having been leached of its copper without complete oxidation of its sulphur.

AT THE INSPIRATION MINE, NEAR MIAMI, ARIZONA, the primary mineralization occurs as an impregnation of schists. The average depth of the leached capping[1] is 367 ft. The average thickness of the enriched ore is 155.5 ft., and its average grade is 2.00 per cent. copper.

AT MORENCI, ARIZONA, the upper limit of the chalcocite zone in the disseminated deposits under Copper Mountain is represented by a curve somewhat less convex than the contour of the surface. The thickness of the chalcocite zone is somewhat over 200 ft., although directly below the summit it reaches 300 ft. in thickness.[2] The depth of leached material in this district varies from a few feet only to over 200 ft. The average grade of the ore now being mined is probably between 2.00 and 2.5 per cent. copper.

Chalcopyrite Enrichments.—Chalcopyritic enrichments are not of so common occurrence as enrichments of chalcocite, but form important deposits in certain districts. Secondary chalcopyrite is commonly associated with bornite, and also with limonite.

AT BINGHAM, UTAH,[3] the enrichment of the Highland Boy orebody is chiefly chalcopyritic. Here the carbonate and oxide ore passes into an enriched zone characterized by chalcopyrite, tarnished and coated with bornite and seamed with limonite: covellite is occasionally an associated mineral. Below this zone of sulphide enrichment the primary ore is found, consisting of lean cupriferous pyrite. The transition between the enriched ore and the overlying oxidized ore is gradual, as is also the change to the underlying primary ore.

AT SAN ANTONIO DE LA HUERTA, SONORA, the primary solution-breccia ore-bodies appear to have received a chalcopyritic

---

[1] Henry Krumb, "First Annual Report of the Inspiration Copper Co,"
[2] Waldemar Lindgren, *P. P.* 43, U. S. G. S., p. 204.
[3] J. M. Boutwell, *P. P.* 38, U. S. G. S., p. 223.

enrichment. The chalcopyrite is here associated with limonite and small quantities of copper carbonates and silicates; the enrichments occur at slight depth below the surface.

The Enrichment of Gold and Silver with Copper.—Under certain conditions, depending upon the presence of proper solvents and the absence of precipitants, gold and silver migrate and form enrichments in company with copper. Surface waters containing chlorides precipitate silver in the oxidized zone as chloride in many instances, and the enrichment of silver with the secondary copper sulphides is thus in part prevented. Not infrequently, however, only a part of the silver is precipitated as residual ore, while the remainder descends with the copper to the water level. The enrichment of gold in copper deposits is likewise variable, and, perhaps, takes place less frequently even than the enrichment of silver, being left in most cases as residual ore in the oxidized zone. Theoretically, silver should be precipitated higher up in the zone of secondary enrichment than copper, as its affinity for sulphur is greater. The copper in enriched copper-silver-gold ores is commonly in the form of chalcocite, although it also occurs in such association as chalcopyrite and enargite.

AT RIO TINTO, SPAIN,[1] the oxidized zone, from 30 to 150 ft. deep, carries from 35 to 50 per cent. iron with traces of copper, arsenic and sulphur. Directly beneath the gossan and resting on the pyritic ore in certain deposits there occurs a bed from 4 to 8 in. thick of earthy, porous material that carries from $10 to $20 gold and about 40 oz. of silver. The upper part of the pyritic zone is enriched by a net-work of seamlets of chalcocite, bornite and chalcopyrite, which minerals occur in progressively less quantity with increasing depth. The pyritic ore directly below the gossan carries from 4 to 5 per cent. copper; at a depth of from 200 to 230 ft. the copper averages about 2 per cent.; at 330 ft. about 1.5 per cent.; at from 425 to 460 ft. about 1 per cent. The primary ore consists of pyrite with a little quartz and chalcopyrite.

AT SILVERTON, COLORADO, the enrichment of silver with cop-

[1] Beck-Weed, "Nature of Ore Deposits," p. 485.

per is well illustrated. According to F. L. Ransome:[1] "The ore first struck, in some cases at the surface, consisted chiefly of argentiferous galena. At a depth varying somewhat in different mines, but which appears commonly to have been less than 200 ft., the galena ore changed to an ore consisting chiefly of highly argentiferous stromeyerite, silver and copper glance. At a still greater depth, commonly at about 500 ft., the stromeyerite changed to argentiferous bornite, still deeper to chalcopyrite and pyrite, and finally to low-grade auriferous and argentiferous pyrite. These changes were more or less irregular and overlapping, pyrite, for example, was found at nearly all levels, and bunches of galena were met with far below the levels at which this mineral ceased to be the dominant ore. Small, rich streaks of bornite were also found, with chalcopyrite and pyrite, below the levels at which it occurred in large masses." According to Schwartz[2] the rich parts of the ore-bodies were in every case associated with open, water-bearing fissures.

The Enrichment of Silver.—The primary ores of silver are readily attacked by oxidation, and in the presence of acid sulphate solutions, yield their silver as soluble sulphate; in this state silver may descend to form secondary enrichments at the water level, where it is readily precipitated by other sulphides. The leaching of the upper parts of silver deposits is similar to the corresponding alteration of copper deposits, with the added factor that a certain proportion of the silver is likely to be precipitated in the oxidized zone as chloride, or as native silver.

The complex sulphur, arsenic and antimony compounds of silver occur as both primary and secondary minerals, and it is often difficult to determine to which process they should be assigned. Where copper occurs with silver in deposits worked over by surface agencies the two metals frequently migrate and precipitate together, although the tendency is for a certain proportion of the silver to be left behind as residual minerals in the oxidized zone and for the secondary sulphide of silver to occur

[1] F. L. Ransome, *Bull.* 182, U. S. G. S., p. 137.
[2] *Trans.*, A. I. M. E., XVIII, p. 144.

near the top of the zone of enrichment. Many of the great bonanzas of silver ores in arid climates are probably secondary enrichments.

Stibnite is of frequent occurrence in silver ores in many districts in Mexico; it is typically a primary mineral, and the manner of intergrowth of the silver minerals with it may indicate their primary or secondary origin.

In the investigation of deposits of silver ores where secondary enrichment is suspected, it should be borne in mind that a relatively small body of rich silver ore may yield important economic return; no such general leaching of rocks or complete alteration of primary sulphides, therefore, should be required upon which to base an expectation of important deposits as is the case with secondary copper ores.

AT THE PROMONTORIO MINE, DURANGO, MEXICO,[1] the primary ore consists chiefly of quartz, galena, and zincblende, a little pyrite, and subordinate chalcopyrite. The oxidized vein-filling consists of quartz, kaolin, hematite, wad, and limonite, with occasional films of malachite, linarite and the remains of sulphides. The minerals that have contributed to secondary enrichment are native silver, chalcocite, and a little chalcopyrite. The secondary enrichments are contained in the oxidized parts of the vein, and in the country rock of the walls and horses. The primary ore-shoots are distinguishable by their comparatively high content of sulphides, by their lack of secondary minerals, and by their habit of being cut off by faults unless occurring in unfaulted parts of the vein. The secondary ore-shoots are recognized by their low content of sulphides, by the presence in their richer parts of the secondary minerals, native silver, chalcocite and chalcopyrite, and by their tendency to follow closely well-defined faults. Primary ore-shoots are dominant in the lower levels and the secondary ore-shoots in the upper levels.

AT GEORGETOWN, COLORADO,[2] the silver lead deposits show evidence of rearrangement by surface agencies. The zone of

[1] F. C. Lincoln, *Trans.*, A. I. M. E., XXXVIII, p. 740.
[2] Spurr and Garrey, *P. P.* 63, U. S. G. S., p. 143.

complete oxidation in these veins usually extends from the surface to a depth of from 5 to 40 ft. only.   The oxidized material is a brown clay; it is, in general, rich ore, containing several hundred ounces of silver.   Below this oxidized material, and mixed with the lower part of it, occur friable, locally powdery, black sulphides and bunches of secondary galena.   These pulverulent sulphides are rich ores also, containing relatively large quantities of silver and of lead, and more gold than occurs at greater depths.   The soft sulphides are found chiefly in cracks and along water courses, and are of secondary origin: they persist to considerable depths below the surface, but in decreasing quantities.   Below the zone of soft secondary sulphides, and irregularly within its lower part, occur secondary polybasite, argentiferous tetrahedrite, and ruby silver, which in quantity diminish steadily but irregularly with increasing depth.   The best ore in most of these veins has been found within 500 ft. of the surface, but locally it extends down to 700 or 800 ft., and in one case to 1000 ft. below the surface.   Much secondary pyrite and siderite occur in the upper parts of these veins.   The primary ore carries from 20 to 30 oz. of silver and the enriched ore from 200 to 300 oz. per ton.   The primary ore consists of galena, zincblende and cupriferous pyrite in a gangue of quartz, and the carbonates of iron, manganese, magnesia and lime.

At Georgetown, Colorado,[1] in the Bismark ore-shoot, the oxidized ores of brown clayey material carry several hundred ounces of silver and extend to a depth of about 40 ft.   Mixed with these ores and extending from 200 to 300 ft. deeper, most prominently along water courses, soft black secondary sulphides are found.   These ores consist chiefly of galena and zincblende, and carry several hundred ounces of silver per ton.   The primary ore is composed of zincblende and galena with some pyrite, in a gangue of quartz, siderite, subordinate barite, and calcite.

At Lake City, Colorado,[2] the primary ore of the silver-bearing fissure veins consists of galena, tetrahedrite, chalcopy-

---

[1] Spurr and Garrey, citing B. B. Lawrence, *P. P.* 63, U. S. G. S., p. 190.
[2] J. D. Irving, *Bull.* 260, U. S. G. S., p. 81.

rite, zincblende and pyrite in a gangue of quartz, rhodonite, rhodochrosite and barite. In the upper parts of the veins secondary ruby silver and argentite are found associated with anglesite, cerussite, limonite and pyrolusite. The values in the upper parts of the veins are commonly high, and bonanzas of ruby silver are found.

AT BROKEN HILL, N. S. W.,[1] the oxidized zone is largely made up of "kaolin ore," which carries oxidized minerals of iron, manganese, lead and copper, and relatively low silver values. In the lower part of the oxidized zone occur the "dry ores," which carry the antimonial and arsenical sulphides of silver, polybasite, stromeyerite, dycrasite, proustite, pyrargyrite, and stephanite. Beneath the dry ores occurs a thin layer, from 3 in. to 6 ft. in thickness, of sooty black sulphides, which carry as much as 250 oz. of silver and 12 per cent. copper. This layer rests upon the primary ores, which consist of an intimate mixture of argentiferous galena and blende, in a gangue of quartz, garnet, rhodonite, and feldspar, with chalcopyrite, arsenopyrite, wulfenite and fluorite as accessory minerals. The primary ores carry from 5 to 36 oz. of silver, 7 to 50 per cent. lead, and 14 to 30 per cent. zinc.

AT NEIHART, MONTANA,[2] superficial alteration is not marked, and no great zones of carbonate or oxidized ores ore found: the deepest general oxidation in the district reaches 170 ft. below the outcrop, but locally oxidation extends as pipes and along drainage fissures to greater depths. The secondarily enriched ores carry polybasite, pyrargyrite and argentite as crusts lining cavities, as thin seamlets through the primary ores, and as sooty sulphides associated with maganese oxides in the oxidized zone. The secondary minerals also occur in the fractures in the shattered country rock. The primary ore contains galena, zincblende, and pyrite in a gangue of quartz and barite.

The Enrichment of Gold.—Gold is readily soluble in the presence of nascent chlorine, which is produced by the action of

---

[1] S. F. Emmons, *Trans.*, A. I. M. E., XXX, p. 204, quoting Jaquet.

[2] W. H. Weed, *Trans.*, A. I. M. E., XXX, p. 435.

sulphuric acid upon chlorides in the presence of the oxide of manganese, all of which compounds are frequently present in the zone of oxidation. Gold is also soluble in solutions of ferric hydrate, sodium sulphide, sodium carbonates and other reagents formed in nature. Gold is precipitated readily by pyritic sulphides and by organic matter, and when in solution with ferric hydrate is precipitated upon the reduction of this compound to the ferrous salt, a change likely to take place at water level. Gold also migrates and precipitates with the antimonial sulphides of silver.[1]

The relatively small bulk of gold as compared with its associated gangue, commonly resistant quartz, tends to protect the gold from solution and also renders it difficult to distinguish a primary gold ore from one of secondary origin. The association of rich ores with post-mineral fissures, the presence of films of oxidized minerals through the ore, and the shape of the ore shoots are the most reliable criteria of secondary origin. Secondary ore-shoots are likely to be longer horizontally than they are deep in deposits of uniform permeability, while primary ore shoots are commonly deeper than they are long.

A majority of gold deposits are larger and richer near the surface than in depth, and their ore-shoots often carry films of oxidized minerals, such as dendritic oxide of manganese, or an incipient oxidation of the associated sulphides, far below the zone of weathering. The high-grade ores found under these conditions are probably the result, in part, of secondary enrichment. Gold is less easily dissolved under average conditions than are most other metals, and is frequently left behind in the upper parts of ore-deposits in which secondary enrichments of other metals have formed at the water level.

It is not uncommon to find rich bodies of gold ore at and just beneath the outcrops of gold veins, due probably to the several processes of mechanical concentration during weathering, residual concentration through the removal of soluble constituents, and also to some extent to solution, migration and enrichment.

[1] W. H. Weed, *Trans.*, A. I. M. E., XXX, p. 432.

AT THE RUBY MINE, MONTANA,[1] an important secondary ore-body, in which the values were about equal in gold and silver, occurred along a post-mineral fault. The gold and the silver appear to have migrated together in antimonial sulphides, notably pyrargyrite, which occurred as coatings on boulders of decomposed rhyolite, and lining cavities in the zone of enrichment.

IN THE GRANITE VEIN, MONTANA,[2] which is described in a preceding paragraph, the low-grade ore of the leached zone carries but little gold; the enriched oxide ore which occurs at the base of the oxidized zone carries from $5 to $16, the underlying enriched sulphide ore from $4 to $8, and the primary ore from $1.50 to $3. The accompanying silver values are concentrated in greater degree than the gold.

AT MONTE CRISTO, WASHINGTON,[3] there is no zone of complete oxidation and the sulphides frequently outcrop. Partial oxidation extends to a depth of perhaps 10 ft. The distribution of the sulphide minerals indicates rearrangement by surface agencies. The upper zone is characterized by galena, gold, and silver, and its ores average $19.00 in gold and 12 oz. of silver. The lower limit of the galena zone follows the contour of the surface at depths of from 100 to 150 ft. Below this occur less regular, but still definite, zones that are characterized respectively by zincblende, chalcopyrite, and arsenopyrite with pyrite. The ores of the last-named zone average $12 gold and 7 oz. of silver.

The Enrichment of Lead.—Lead is less easily soluble than most other metals: upon oxidation it alters first to the sulphate, anglesite, and then to the carbonate, cerussite. The carbonate, apparently, may also form directly from the sulphide. Owing to the relative insolubility of these minerals lead is dissolved slowly and is readily precipitated and the process of enrichment

---

[1] W. H. Weed, *Trans.*, A. I. M. E., XXX, p. 433.
[2] W. H. Emmons, *Bull.* 315, U. S. G. S., p. 39.
[3] J. E. Spurr, *Twenty-second Annual Report*, U. S. G. S., and "Geology Applied to Mining," p. 280.

is a slow one. Long continued action of surface waters, however, with repeated solution and precipitation may produce a zone of lead enrichment higher up in a deposit than the enrichments of associated metals such as copper and zinc, whose secondary sulphides are commonly found at the water level. Galena is slightly soluble in water, and in solutions containing sodium sulphide. The sulphate of lead, anglesite, is slightly soluble in water, and the carbonate, cerussite, is soluble in carbonated waters. Galena is precipitated by hydrogen sulphide produced by the reactions of solutions upon sulphides, by organic matter, and probably when migrating as the sulphide, by direct replacement of calcite.

While in most deposits the upper lead bearing zone is in large part made up of residual galena, anglesite and cerussite, migration and enrichment have probably assisted in concentrating these minerals. The migration of lead in the presence of acid sulphate solutions is apparently more difficult and less complete than in certain limestone deposits where carbonated waters have probably acted as the solvents.

The examples of secondary enrichments of galena are neither numerous nor clear, the usual occurrence of lead minerals in deposits worked over by surface agencies being as residual minerals.

At MONTE CRISTO, WASHINGTON, as described in a preceding paragraph, the upper part of the sulphide zone is characterized by galena, the lower limit of which follows roughly the contour of the surface, indicating a concentration by surface agencies.

At GEORGETOWN, COLORADO,[1] in the veins of the Freeland group galena was more abundant in the upper few hundred feet, while but small quantities of this mineral is contained by the characteristic pyritic ore. In the zone of relatively abundant galena, brown carbonates (siderite, rhodochrosite, and some barite), were abundant, as was also tetrahedrite, which occurred in cracks through the older sulphides. Both the carbonates and tetrahedrite were in places clearly of secondary origin.

[1] Spurr and Garrey, *P. P.* 63, U. S. G. S., p. 149.

IN THE UPPER MISSISSIPPI VALLEY [1] deposits of zincblende and galena associated with marcasite, pyrite and subordinate chalcopyrite occur in unaltered, flat lying dolomites and limestones far from any known igneous rocks, and apparently, in most cases, unconnected with any important fissure. In form these ore-bodies occur in vertical crevices, in irregular ore-bodies at the juncture of an ore bearing horizon with a vertical fissure, in pitches and flats, and in thin, flat lying disseminations. The ore-bodies are closely related to certain very shallow structural basins, and to certain shale beds known locally as oil rock,

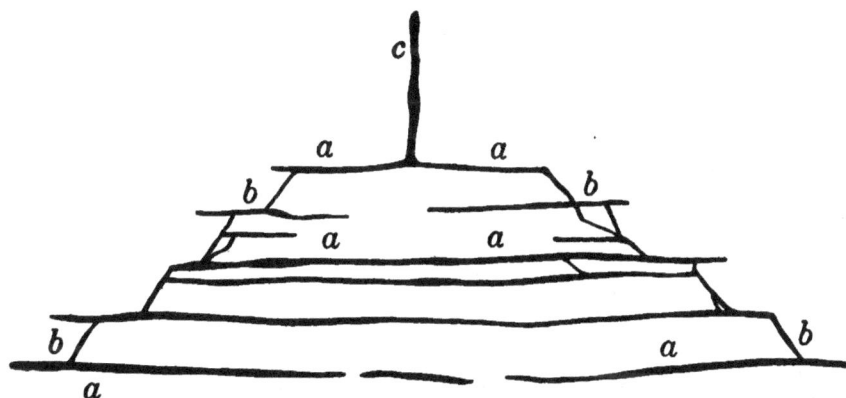

FIG. 75.—Diagram illustrating type of "pitch and flat" deposit of the Upper Mississippi Valley. *a*, Flats; *b*, pitches; *c*, vertical crevice. *After Grant.*

which contain a large percentage of fossil gum. These ore-bodies are believed by many to represent concentrations by descending waters of sparsely disseminated mineralizations; their genesis must be considered doubtful. The common vertical distribution of the several ore minerals is: galena in the uppermost zone, below which is found zinc carbonate ore, underlain in turn by zincblende ores below the water level, while the lowest zone is pyritic; the several zones commonly merge into each other. The localization of the ores appears to be connected with the water level; lead ores occur above the water level, zinc carbonate from slightly above to slightly below, and the zincblende below but near to the water level. A secondary distribution is thus indicated.

[1] H. F. Bain, *Bull.* 294, U. S. G. S., p. 46.

While scanty occurrences of zincblende and galena are known in both the overlying and underlying rocks, the important mineralizations are confined to the Galena dolomite, and in those districts where this rock has been largely eroded they also occur in the underlying Platteville formation: while ore-deposits are found in all horizons of these beds certain strata are locally more likely to contain ore than others. The ore-deposits are in the main confined to very shallow synclines. The aforementioned "oil rock" is a brown or black shale varying in thickness from 6 to 8 ft. down to a few inches; it contains thin lenses of dolomite and the shaly material commonly does not form single bands more than one foot in thickness. This bed contains fossil gum that appears to have been active in precipitating the ore minerals. Pitches and flats are commonly developed above the oil rock, which itself frequently contains disseminated ores. The oil rock is seldom seen in any considerable thickness outside of the mines, though it is not of such a nature as to be especially subject to weathering.

The Enrichment of Zinc.—Zincblende is readily attacked by oxidizing solutions and the resulting sulphate is very soluble. Zinc is usually completely leached from the oxidized zones of most deposits, but sulphide enrichments in igneous rocks are comparatively rare; in sedimentary rocks, however, migrating solutions carrying zinc frequently replace certain beds of the country rock with oxidized zinc minerals. Such deposits, even when of important extent, are difficult to detect, as the replacement commonly retains the structure and to some extent the color of the replaced rock, while the low specific gravity of these minerals does not call attention to their presence. Zincblende, under certain conditions, apparently migrates with ease in limestone.

In the Upper Mississippi Valley secondary enrichments of zincblende are common in the deposits in limestone and dolomite. The zincblende occurs just below the water level and is usually associated with marcasite and some galena. This zone passes in depth into ores that carry abundant marcasite with but little zinc.

At Leadville, Colorado, the zinc is commonly leached from the oxidized ores near the surface and forms large replacement deposits of oxidized minerals in depth, the existence of which was not suspected for many years.

At Pachuca, Mexico,[1] the silver veins are impoverished in depth by large quantities of barren zincblende.

The Enrichment of the Lesser Metals.—Under certain conditions, nickel, cobalt, arsenic, antimony, tin and cadmium appear to be capable of solution, migration and enrichment.

The Migration of Gangue Minerals.—The solution, migration, and occasionally the precipitation, of gangue minerals are important processes in the rearrangement of ore-deposits by surface agencies. The residual concentration of gold in the oxidized zone is the result of the solution and removal of associated gangue minerals and sulphides.

The carbonates and alumina appear to be especially susceptible to solution and removal, while under certain conditions quartz is a mobile mineral.

The oxides of manganese are apparently soluble in solutions of ferrous sulphate;[2] the presence of dendritic oxide of manganese in many deep deposits is probably the result of the migration of this compound in solution.

[1] Waldemar Lindgren, *Trans.*, A. I. M. E., XXX, p. 650.
[2] F. W. Clarke, quoting F. P. Dunnington, *Bull.* 330, U. S. G. S., p. 458.

CHAPTER XI

OUTCROPS

In the examination of an undeveloped prospect a decision must be arrived at from an inspection of the outcrops and the exposures in a few shallow pits. Prospects that are offered for sale rarely expose any important quantity of payable ore, work having usually been stopped when the immediate exploration no longer yielded favorable results. It should be borne in mind, furthermore, that a great majority of prospects have been examined many times, and if of conspicuous promise, would have been acquired for development. In most cases the problem for the engineer, therefore, is to discover the traces of a valuable mineralization that has disappeared through solution, or the conditions that indicate the possibility, or probability, of underlying secondary enrichments.

The available data being meager, every feature should be the subject of careful study—the condition of the outcrop, the structural relationships, the associations of minerals, the alterations of wall rocks, the chances for underlying enrichments, and so forth, as well as the actual assay value of the material exposed.

It is frequently advisable to spend the time and money necessary to have trenches dug at various significant points along a promising outcrop; such exposures, even if the depth attained is slight, often disclose conditions that are not apparent at the actual surface, as for example, the existence of soft minerals and the distribution of the various minerals through the mass of the deposit. Trenches also permit samples to be taken from points that were not accessible during former examinations.

The Relation between Length of Outcrop and Persistency of Vein in Depth.—Strong, persistent outcrops of uniform width may be taken to indicate the probable size and character of the underlying vein, whose persistency in depth is likely to be proportional to the length of its outcrop. Fissures that may be traced for long distances on the surface are commonly found to be equally persistent in depth, while short, branching and irregular outcrops are usually indicative of similar irregularities in the underlying deposits. Outcrops that comprise a series of large irregular masses, perhaps connected by narrower veins, are in general likely to become smaller in depth.

The Relation between Size of Outcrop and Width of Vein in Depth.—In a vein or deposit that is harder and more resistant than the enclosing rocks, erosion is likely to be halted at the widest part, that part offering the greatest resistance to erosion and the greatest protection to the adjoining rocks. Such outcrops, therefore, are likely to be larger than the average of the underlying deposit. The converse of this condition is also true. A vein or deposit that is softer and less resistant to erosion than the enclosing rocks is likely to become larger in depth. Erosion tends to remove the surface of such a deposit faster than the adjoining wall rocks, forming a depression, the walls of which tend to close together upon the removal of the intervening soft material; occasionally, a narrow and inconspicuous gouge-filled fissure is the only surface indication of an important vein of soft minerals.

Brecciation and Post-mineral Fracturing.—In disseminated deposits the most intense primary mineralization is likely to be connected with a thorough brecciation, as are also disseminated enrichments. The study of post-mineral fracturing, therefore, is of as much importance in the investigation of these deposits as is the study of the minerals of the outcrops themselves.

Meandering of Outcrops on Hillsides.—The outcrop of a vertical vein is a straight line, whatever the slope of the surface, and the outcrop of a horizontal bed follows a contour around all hillsides. A vein of intermediate dip, however, outcrops to the right or to the left of its strike at any horizon in accordance with its dip and the slopes of the surface. An idea may be obtained in the field as to the dip of a vein by paralleling a book or other plane surface with several parts of the outcrop at different elevations, while in mapping, simple trigonometric calculations or graphic solutions will give the heights or horizontal positions of the vein at any desired point: contouring is expensive and is usually unnecessary.

Down Hill Creep.—An outcrop situated on a steep hillside is likely to overturn in the direction of the slope of the hill; loose fragments of the partially disintegrated outcrop become

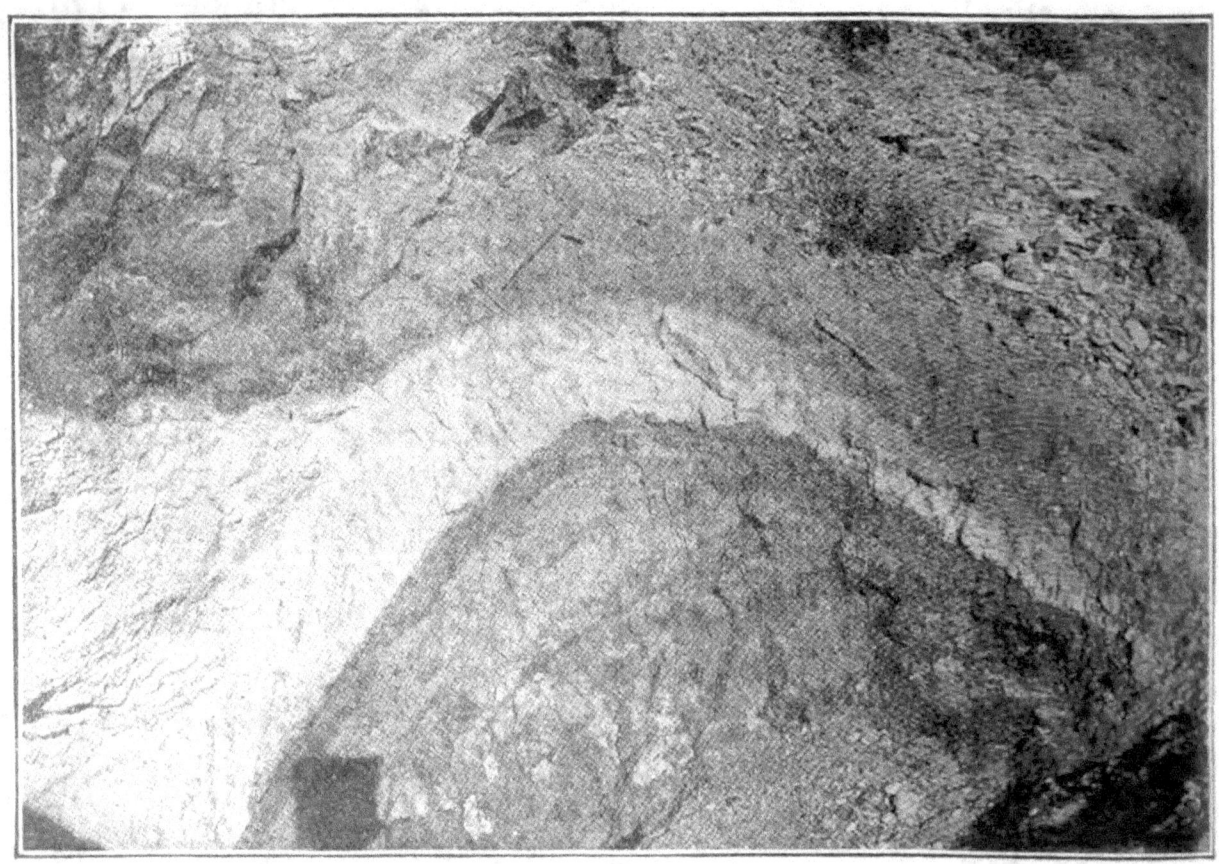

Fig. 76.—Outcrop of a vein at Bullfrog, Nevada, showing surface overturning and down-hill creep. *After Ransome.*

mingled with the surface material, and when at some distance from the parent mass are widely separated and constitute "float." The apparent dip of a vein that outcrops along a steep hillside is, therefore, likely to be flatter than its true dip below the mechanical influence of erosion.

The Topographic Expression of Mineralization.—Mineralization occasionally finds local expression in the topography.

Silicified areas and resistant outcrops may form prominent ridges or knobs, and soft or altered rocks may result in depressions or saddles in the ridges. Faults are occasionally prominent topographic features where one wall is notably more resistant to

Fig. 77.—Fault plane developed into a scarp by erosion, Globe, Arizona; the rock on the left (hanging-wall) is quartzite, that on the right (foot-wall) is granite. *After Ransome.*

erosion than the other, but, in general, unless of great displacement, faults are not likely to be represented in the topography. Different kinds of rocks upon weathering produce characteristic topographic outlines, as, for example, the plateau-like hills and steep talus slopes of horizontal sedimentary beds, or the ragged outlines common in areas of volcanic rocks. While topographic relief is rarely significant in the examination of mining properties, it may be a valuable guide to the prospector, as many mineralized areas are connected with low, rounded foot hills of notably different outline from the general relief of the region.

Outcrops of Deposits Formed at Slight Depth.—Deposits formed at slight depth below the surface are characteristically irregular in their upper part, and tend to consolidate and to become structurally more regular with greater depth.

Porosity of Outcrops.—The outcrops of ore-deposits that have yielded to oxidation are commonly porous and cellular, owing to the removal of certain of the original constituents in solution. Such outcrops are favorable in that they indicate solution and possible secondary enrichments of the dissolved substances, or, at least, the original presence of easily dissolved minerals which in their unaltered state may have been valuable. A massive, tight outcrop, on the other hand, is usually indicative at slight depth of the typical value of the deposit, which has presumably been but little affected by secondary agencies.

Casts in Resistant Gangue Minerals.—Upon the oxidation and leaching of an ore composed of sulphide minerals in a resistant gangue, the outcrop is likely to retain the casts, or open spaces, left by the removal of the sulphides. In many instances outcrops exhibit abundant casts of minerals that have completely disappeared, and whose presence originally would not be suspected without a close examination for such evidence.

The crystallization of pyrite appears to be interrupted when it contains a notable proportion of copper. Pure, barren, pyrite commonly crystallizes as cubes or other isometric forms, while chalcopyrite is commonly distributed in an irregular manner; the crystallization of cupriferous pyrite, in general, is wavy and irregular, although occasional cubes may be noted. A less strongly emphasized but still marked relation exists between pure, barren galena and highly argentiferous galena. This mineral is less likely to be well crystallized when it carries an important proportion of silver. An examination of the casts of these sulphides in a leached outcrop, therefore, is likely to yield evidence to some extent in regard to the value as well as to the kinds of the sulphides originally present.

The Composition of Outcrops.—The minerals present in outcrops are those primary constituents of the ore most resistant to solution, and the most resistant of the products of alteration. The most resistant of the primary minerals is usually quartz, which commonly forms a large proportion of the outcrops of the deposits in which it is an important constituent. Magnetite and specularite are also relatively resistant minerals. Of the products of oxidation and hydration kaolin and limonite are the most resistant, and are commonly found in or close beneath the outcrop. Sericite likewise is resistant, but is likely to be changed to kaolin where sulphuric acid solutions are prominent in accomplishing the surface alteration.

The oxides of manganese are resistant, and are frequently found in large quantities in the oxidized parts of ore-deposits, even those in which manganese forms a quite subordinate proportion of the primary ore.

Gold is a resistant mineral, and is frequently concentrated at the surface and in the oxidized zone, as is discussed in a preceding paragraph.

The more resistant of the ore minerals are galena and its oxidation products, anglesite and cerussite, chloride of silver, native silver, cuprite, native copper, chrysocolla and to a lesser degree the carbonates of copper, all of which minerals are likely to be left behind as residual ores during solution and migration.

In copper deposits that have been subjected to thorough leaching and alteration, the outcrops are likely to consist of soft white kaolin and quartz, carrying at the surface, perhaps, trifling quantities of limonite, manganese oxide, and chrysocolla as stains. Such outcrops are most favorable for the existence of chalcocite enrichments in depth.

The thorough alteration of an intensely mineralized mass may remove all traces of the primary ore minerals, while adjacent rocks that received a scanty mineralization only and so remain relatively unaltered, often retain traces of the primary minerals and so furnish a clue to the original nature of the principal deposit.

The Oxides of Iron in Gossan.—The oxides of iron, either massive or as stain, are frequently the most prominent constituents of outcrops and of the superficial parts of ore-deposits that have suffered oxidation. The condition and occurrence of these minerals are often indicative of the character of the mineralization in depth.

The final product of the oxidation and hydration of iron minerals is limonite, and the other oxides of iron on thorough alteration yield this mineral. Magnetite in an outcrop is commonly present as an unaltered primary mineral, and may not be taken to represent the alteration product of a sulphide mineralization. Specularite, of characteristic crystal form, is likewise a primary mineral, and not the result of the alteration of sulphides: micaceous hematite should be distinguished from specularite, as it is frequently found as the alteration product of sulphides, the micaceous structure having been developed by stress exerted after its formation. Certain gangue minerals, such as lime-iron garnet, yield limonite upon oxidation; in most cases limonite of this origin occurs as soft earthy masses, mixed with unaltered contact minerals as, for example, partially decomposed epidote, and is commonly distinguishable from limonite resulting from the oxidation of sulphides.

In many outcrops in arid regions whose underlying ore deposits have been explored, the limonite resulting from the oxidation of pyrite occurs as a massive brown mineral, while the chalcopyrite of the original ore is represented by seamlets and patches of soft, bright red hematite.

The Condition of Outcrops Indicative of Secondary Enrichments in Depth.—Features of outcrops that indicate the possibility of enrichments in depth are the residual indications of a good primary mineralization together with a porous or brecciated structure, or the presence of post-mineral fractures.

A majority of enrichments of secondary sulphides are probably due to the migration of sulphate and sulphuric acid solutions; these solutions frequently leave traces through the presence of a kaolinitic alteration of the associated rocks, or of kaolin in the outcrop or oxidized zone. Resistant minerals that remain after a thorough alteration of this kind are kaolin, limonite and quartz. Feldspars are completely altered and sulphides are absent. The presence of unaltered feldspars or of most of the other usual gangue minerals except quartz indicates a partial alteration at best, and the presence of sulphides indicates an incomplete solution. Galena is a resistant mineral as compared with other sulphides, and not infrequently overlies important enrichments of other metals, but the presence of unaltered pyrite, chalcopyrite or zincblende renders it unlikely that important enrichments will be found through deeper exploration.

The outcrops of suspected disseminated chalcocite enrichments should contain little else than kaolin, limonite and quartz, together with, perhaps, some unaltered sericite, itself a product of primary altering agencies. The presence of pyrite in such an outcrop or in the oxidized zone is a most unfavorable sign; cases are known where pyrite was found above secondary chalcocite enrichments, but in every case it was partly altered, crumbly, and was contained within thoroughly kaolinized rock, the supposition being that the copper was leached from the pyrite under conditions that did not permit a complete oxidation of the sulphur. Where the containing rock shows the outlines of the feldspars the kaolinization must be considered unsatisfactory.

Efflorescences of soluble salts in the outcrops or at slight depth below them are frequently instructive. An efflorescence of copper sulphate is indicative of the solution of chalcocite, and may or may not be mixed with iron sulphate. An efflorescence of iron sulphate with but little copper is commonly indicative of pyrite at no great depth. Sulphur and a yellowish-green sulphate of iron are probably not formed except close to oxidizing pyritic sulphides. Other efflorescences frequently found are alum and zinc sulphate, the latter characteristic of primary ores below any zone of chalcocite enrichment; complex sulphates of aluminum and other bases are also formed in the leached zone.

Where large quantities of pyritic sulphides have oxidized and have in part been removed in solution it is not uncommon to find the conglomerates of stream beds cemented by limonite, and, in some instances, carrying oxidized copper minerals; the presence of such conglomerates may be taken to indicate conditions under which migration and enrichment may have taken place within the deposits themselves.

The elevation of an outcrop as compared with the drainage level, or the elevations of the known enrichments of the district, are important factors in the consideration of possible secondary enrichments.

Disseminated chalcocite enrichments appear to be confined to regions of slight rainfall: important enrichments of copper and other metals are found in veins that have suffered post-mineral fracturing under all climatic conditions, except where rapid erosion or glaciation continuously exposes fresh surfaces of primary ores and thus does not permit the operation of surface agencies.

Rock Alteration as a Guide to Ore-Deposits.—The several types of rock alteration are discussed in a preceding chapter where their close relationship with mineralizing processes is emphasized. In the field it is usually possible to distinguish between a primary or hydrothermal alteration of the rocks and the results of ordinary weathering, and also between either of these and kaolinization, but where such distinction is doubtful, slides should be cut and examined under the microscope, when the type of alteration will become apparent. A "highly altered condition" means nothing unless its type and probable relation to mineralization are understood.

The Outcrops of Kaolinized Rocks.—Upon the kaolinitic alteration of rocks carrying pyritic mineralizations the products of the alteration are likely to be similar whatever the original nature of the individual rocks. On Shannon Mountain, Arizona, granite, porphyry, and shales have all suffered intense kaolinitic alteration, and the resulting mass of kaolin, sericite, and quartz with associated limonite and chalcocite may rarely be differentiated in the field into parts representing the original types of rocks. In granitic rocks that were not sericitized before suffering kaolinization the quartz phenocrysts remain clearly distinguishable, and in hand specimens tend to obscure the degree of alteration; a kaolinitic alteration, however, that has not obliterated the outlines of the feldspars cannot be considered thorough, and outcrops of this nature are rarely underlain by important enrichments. Thorough leaching and kaolinization usually removes the iron as well as the other bases, and the dumps of workings in the leached zones are commonly white in color, in sharp contrast with the brown or red color of the surface. In deposits that originally carried abundant pyrite, much limonite may remain in the leached zone. The line of demarkation between the leached and the enriched zones in such cases is well marked, the former being stained by iron, while the latter is white in color.

The Outcrops of Contact Deposits.—The minerals developed by contact metamorphism are, in general, resistant to oxidation and erosion, and are likely, therefore, to form conspicuous outcrops. The tightness of these minerals and the slowness with which they decompose under the influence of surface agencies protect the sulphides contained within them and so in large measure prevent migration and enrichment. A further hinderance to secondary enrichment in deposits of this nature is the usual presence of active precipitants. As a result of these conditions secondary enrichments are rare in contact deposits except where the sulphides were present in large masses; in all other cases the outcrops of contact deposits are likely to be indicative of the values contained by the deposit at all depths.

Deposits of Surface Origin.—Certain deposits of surface origin present close imitations of the outcrops of ore-deposits, but are not underlain by valuable minerals. Bog iron ore (limonite) is a familiar example of surficial deposit not connected with any underlying mineralization, and many deposits of the oxides of manganese also occur in like manner. Nodular concretions of manganese oxides that form on the sea bottom are likely to be concentrated into such superficial deposits, as is also, upon erosion, the manganese contained in small quantities by many rocks. Bog iron ores commonly contain casts of vegetable remains, sand, and silt, and may also be recognized by their bedded form. Surficial deposits of limonite and manganese oxides may generally be distinguished from the outcrops of mineral deposits through their lack of associated minerals characteristic of outcrops. Furthermore, after slight exploration their structure is revealed and their superficial nature becomes apparent. Pyritic deposits are occasionally met with where the sulphides have replaced roots, or other organic matter, and whose superficial formation is evident.

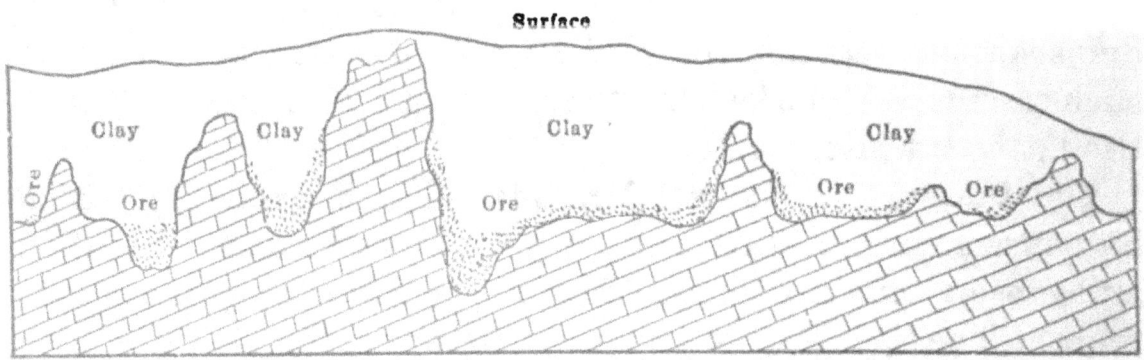

Fig. 78.—Section of open cut, Bertha Zinc Mines, Virginia, showing the superficial position of the residual ore. *After Watson.*

Microscopic Examination of Specimens.—The investigation of an outcrop is largely a search for residual conditions indicating that an important mineralization has been removed from the surface by oxidation and solution. It is often advisable to have slides cut from specimens of an outcrop and its associated rocks, and to have them examined under the microscope. Information is thus gained of the character and degree of the alteration, and in many cases of the mineralogical nature of the original ore and of the distribution of the several minerals. Specimens taken for this purpose should be chosen carefully, and a record of them kept in the same way as is the custom with samples taken for assay: furthermore, it is best always to reserve a specimen, preferably part of the same piece that is sent away, for purposes of comparison upon receipt of the results of the examination. Microscopic examination is of great value, if only for corroboration of the evidence gathered in the field, but may also yield information and indicate possibilities that without it would not be suspected.

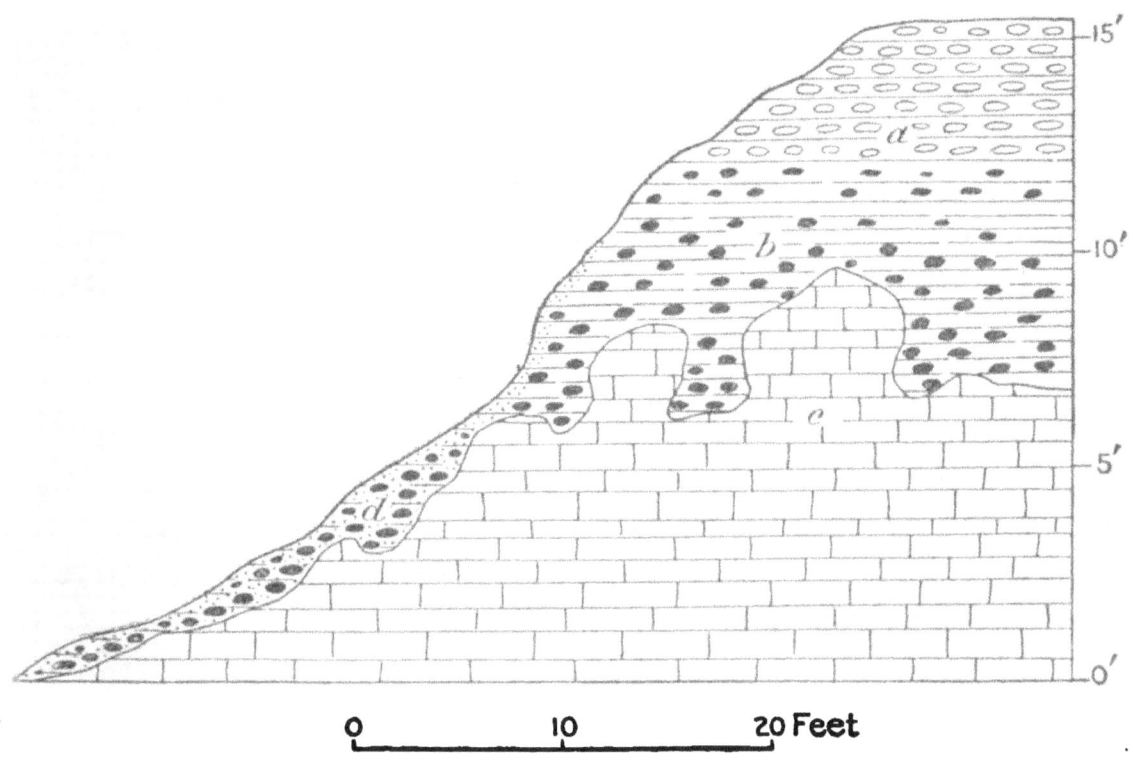

FIG. 79.—Example of surficial deposit; nodules of manganese oxide. *After Harder.*

DESCRIPTIONS OF OUTCROPS

AT NACOZARI, SONORA, MEXICO, the outcrops of the Pilares solution-breccia are prominent. The country rocks of this district are andesitic and rhyolitic porphyries whose surface over an area of several square miles is stained red, yellow or brown by the oxides of iron resulting from the oxidation of a disseminated cupriferous pyritic mineralization. Shallow leaching and attendant unimportant enrichments of disseminated chalcocite and chalcopyrite are present at many points in the district, but the depth and degree of alteration and enrichment are apparently too slight to yield ore-bodies of commercial importance. The walls of the gulches frequently show efflorescences of chalcanthite, usually associated with ferric sulphate and sulphur, the ear marks of a near-by unaltered pyritic mineralization: a white, probably aluminous, precipitate is present at many points where solutions are oozing out of the rocks, and this, apparently, is also a sign of shallow alteration. In the vicinity of the Pilares ore-body, the only important copper deposit in the district, the rock is strongly sericitized, and in this respect differs from the remainder of the district seen by the writer. The outcrop of this ore-deposit is composed of a striped breccia, seamlets of ore minerals being separated by long, thin slabs or sharp angular fragments of the country rock that frequently show a curving parallel arrangement of ore- and rock-areas.[1]

The chief ore mineral is chalcopyrite, with some bornite, associated with pyrite and quartz. In the outcrop, the seamlets residual after the ore are made up principally of micaceous hematite and quartz, with small particles of copper carbonates. Through the brown micaceous hematite, and as individual seams, occurs bright red hematite, in distribution following that of the chalcopyrite in the ore of the sulphide zone, and evidently residual after this mineral. The outcrop is of small extent as compared with that of the deposit as developed in the deep levels.

At Jerome, Arizona, the conditions exposed in the open pit of the United Verde mine are indicative of intense mineralization. Here the principal country rocks appear to be schist and a

---

[1] This structure is considered by Mr. S. F. Emmons to indicate mineralization attended by a splitting off of the country rock in concentric shells under the influence of the mineralizing solutions; this type of deposit is termed by him a "solution breccia." The writer has seen the same type of mineralization at several other districts, all of them in Sonora; in every case deposits were contained in deep flows, or sills, of andesitic or rhyolitic porphyry. At San Antonio de la Huerta the copper deposits are of this type; at La Trinidad, Sahuaripa, the lead silver deposits occur in a solution breccia, as do also the silver deposits at Guadalupe; the gold deposits near Bacoachi are said to be solution breccias.

ather basic granite-porphyry that carries a heavy pyritic mineralization. The residual minerals remaining in the walls of the pit from which the surface ores were mined are white kaolin, quartz, limonite, and oxides of manganese in which occur copper carbonates and silicates. The outcrop is said to have carried gold values. Sharp lines of demarkation exist between bands of unaltered schist and the mineralized and altered parts of the deposit. Kaolinization does not appear to have affected the wall rocks. The principal ore minerals in depth are auriferous chalcopyrite and bornite.

At Morenci and Metcalf, Arizona, the lode and the disseminated deposits in porphyry have inconspicuous outcrops. The lodes are represented at the surface by quartz-filled fractures through sericitized and kaolinized porphyry. A light iron stain is common at the surface throughout the intrusive areas, but, in general, the iron as well as the copper has been removed from the mass of the leached zone by solution. Except where the secondary ores actually outcropped, as is said to have been the case with the Metcalf ore-bodies, there is but little indication of copper at the surface, although a little chrysocolla is occasionally seen, and light efflorescenses of chalcanthite are common on the walls of shallow workings driven into the porphyry. The most promising areas for disseminated chalcocite enrichments are indicated by abundant casts, residual after an intense primary pyritic mineralization, together with ramifying quartz veinlets associated with a thorough kaolinization of the rock. All of the important enrichments in this district occur high up in the hills; the canyons are in every case in lean primary ore, and extensive exploration of the lower hill slopes has been without result.

At Bisbee, Arizona,[1] one only of the large pyritic ore bodies outcropped at the surface; in this deposit the usual oxidized copper minerals were present in quantities sufficient to constitute ore. The limestone in the vicinity of many of the ore-bodies is fractured and carries disseminated pyrite. The rusty

[1] F. L. Ransome, *P. P.* 21, U. S. G. S.,

14

outcrops of this pyritic mineralization have in some instances indicated the presence of underlying ore-bodies; slight depressions were noted at the surface above certain ore-bodies, probably due to subsidence upon the removal in solution of certain constituents of the deposits. Many of the most important of the ore-bodies of this distinct, however, are not represented in any way at the surface. The intensely mineralized porphyry of Sacramento Hill taken in connection with the fracturing, silicification and marmorization of the limestone are surface conditions patently favorable to the existence of ore-deposits.

At Bingham, Utah,[1] the outcrops of the lead-silver, and the copper deposits of the Old Jordan group are inconspicuous. The containing limestone is irregularly altered to white or gray marble, often carrying structureless patches of decomposed powdery limestone; silicification is extensive, quartz occurring as granular masses, as fine compact replacements, and as cherty bands and patches. The outcrops of the lodes, which are traceable with difficulty, are belts of shattered and discolored rock; the fissure fillings are marked by crushing and are stained by the oxides of iron and copper carbonates. The replacement deposits are represented at the surface by shattering, silicification, and discoloration of the limestone. At many points the rusty, siliceous outcrops carried gold values. An upper zone of carbonates and oxidized ores here is commonly underlain by enriched sulphides.

At Park City, Utah, the ores occur as lode and replacement deposits in limestone; outcrops are scanty or lacking. Marmorization and silicification of the limestone are present in places, and shattering and discolorization are the chief surface indications of the important deposits.

In the Coeur D'Alene District, Idaho,[2] the outcrops of the important lead-silver deposits in limestone are inconspicuous. Soil and vegetation cover the hillsides, and the material of the lodes is neither so superior to the enclosing rock in hardness or

[1] J. M. Boutwell, *P. P.* 38, U. S. G. S., p. 238.
[2] F. L. Ransome, *P. P.* 62, U. S. G. S., p. 129.

durability as to form bold outcrops, nor so easily eroded as to produce trenches or saddles in the topography. The courses of the lodes may not be followed at the surface without the aid of test pits or trenches. The ores at the surface carry galena with a little cerussite, limonite and copper carbonates.

AT METCALF, ARIZONA, large contact metamorphic deposits form the summit of Shannon Mountain, their resistant minerals having halted erosion at the horizon of greatest development. The limestone strata have been largely altered by contact metamorphism to magnetite and garnet with abundant pyrite and associated chalcopyrite.

Oxidation has worked over these deposits thoroughly; while part of the copper has been removed by solution, in the sediments most of the copper originally present probably remains as residual ores; the magnetite and garnet have been partly altered to limonite and quartz. The primary pyritic mineralization was sufficiently intense to have permitted access by surface waters and thorough oxidation. The residual ores are chiefly azurite, malachite, chrysocolla, and brochantite, associated with limonite, hematite, and the oxides of manganese. In the porphyry areas the surface is kaolinized and leached, and underlying enrichments of chalcocite are found in large deposits, both as disseminated enrichments, and in lode-like deposits along fissures. Near the top of the enriched zone further oxidation has produced cuprite and some native copper from the chalcocite. The shales, in part altered to hornfels by contact metamorphism, received in places a disseminated mineralization of pyrite and chalcopyrite which have probably been somewhat enriched by circulating solutions.

AT SAN PEDRO, NEW MEXICO,[1] copper occurs as chalcopyrite associated with garnet in extensive contact deposits. These deposits are confined to the lower part of a laccolithic roof of limestones. The shaly limestone is in places altered to hornfels and carries tremolite and diopside as coarse crystals: this type of altered rock carries no ore; above these beds occur garnetized

[1] Waldemar Lindgren, *P. P.* 68, U. S. G. S., p. 173.

beds from 50 to 100 ft. thick and of great horizontal extent, through which the ore is irregularly distributed. The primary ore is composed of chalcopyrite, with some gold, in a yellowish-garnet, chiefly andradite, calcite, and lesser quantities of tremolite and wollastonite. Oxidation has had but slight effect on these deposits, in which apparently there has been no migration or enrichment.

AT VIRGINIA CITY, NEVADA, the Comstock Lode outcrops for a long distance as siliceous masses and quartz veinlets cementing a fractured zone. This lode, unlike most deposits, occupies a fault of large displacement. The country rocks are propylitized andesites. The lode is continuous over its central part, but branches at either end; while broad and scattered at the surface, it becomes more regular in depth. The principal values at the surface were as chloride of silver. The bonanza ores contained stephanite, polybasite, argentite, and native gold, associated with small quantities of galena and zincblende in a quartz gangue.

AT TONOPAH, NEVADA[1] the important veins showed prominent and continuous outcrops of white quartz; the first samples broken from the veins, although rich, showed no ore minerals, and were of unpromising appearance. The quartz has in places a purplish color, due to minute particles of argentite. The ores near the surface carried chloride, bromide and iodide of silver associated with limonite and oxides of manganese. The country rock is a sericitized andesite.

AT SILVERTON, COLORADO,[2] the outcrops of the stock deposits of the Red Mountain area form prominent siliceous knobs composed of silicified andesite carrying finely disseminated pyrite, sericite and kaolin. These knobs are thoroughly fractured and contain vugs, cavities, and ramifying caves. The residual ores were found chiefly in the caves, as beds of sandy or clayey material, and on their walls associated with porous, spongy masses of quartz. The oxidized ores carried argentiferous cerussite and anglesite associated with siderite, barite, "oxide of iron" and

[1] J. E. Spurr, *P. P.* 42, U. S. G. S., p. 122.
[2] F. L. Ransome, *Bull.* 182, U. S. G. S., p. 233.

kaolin. At slight depth argentiferous galena formed the principal ore; this gave way in depth to argentiferous enargite, chalcocite, bornite and chalcopyrite, which in turn were underlain by lean primary pyritic sulphides.

AT CRIPPLE CREEK, COLORADO, the telluride gold ores occur with a very scanty gangue, and the outcrops of the veins are not conspicuous. "As elements of geological structure, the lode fissures at Cripple Creek are exceedingly inconspicuous. They are marked neither by bold outcrops of quartz nor by superficial bands of ferruginous gossan. They seldom fault perceptibly the structures that they traverse, and are not sufficiently different from the enclosing rocks as regards resistance to erosion, to have influenced perceptibly the topographic development of the district. It is this obscurity that retarded the discovery of the ore-deposits, and that to-day renders it impossible to follow the veins over the surface without first stripping off the soil and loose rock, or sinking test pits."[1]

The surface ores here contain dull native gold associated with tellurites in ferruginous clays, kaolin and alunite. Hydrothermal metamorphism was an inconspicuous process in this district.

IN THE BLACK HILLS, SOUTH DAKOTA,[2] the large Homestake ore-bodies outcropped as iron stained chloritic slates, quartz, and porphyry, which in places carried as high as $16 in gold. These altered gold-bearing rocks were not sufficiently resistant to erosion to produce prominent outcrops. In the open cuts it is not possible for the unpractised eye to distinguish the slates that are sufficiently mineralized to constitute ore from those that are practically barren. In general, the rocks within the ore zone appear to be somewhat more completely silicified, and to carry more iron stain than the country rocks, and also to have been subject in greater degree to fracturing and folding. In the surface ores the quartz occurs in three phases: as thin silicious layers intercalated in the slates, as thin seams that frequently follow the lamination and bedding, but occasionally cut across

[1] Waldemar Lindgren and S. F. Emmons, *P. P.* 54, U. S. G. S., p. 153.
[2] J. D. Irving, *P. P.* 26, U. S. G. S., p. 57.

them, and also as veinlets independent of the structural features of the containing rock, which last are the most important. The gold in these deposits occurs in finely disseminated particles, and appears to be of later origin than the quartz and rock gangue.

In the Mogollones District, New Mexico,[1] the country rocks are a series of flows, fragmental beds and tuffs, composed of soda-rhyolite, andesite and basalt. The rocks in the mineralized areas exhibit propylitic alteration; silicification, resulting in a greenish-gray hornstone, is prominent along the veins. Sericitic alteration is absent. The veins are partly filled fissures and partly the result of replacement of the walls along fractured zones. The topography is rugged, and the veins being harder than the altered wall rocks, form bold outcrops. The veins carry native silver, argentite, chalcocite, pyrite, chalcopyrite, and bornite associated with specularite in a gangue of quartz, calcite and fluorite. In certain of the veins in which copper is present in small quantity the ore minerals are finely disseminated through the gangue. In another type of vein copper forms a large proportion of the total value of the ore and sulphides are abundant. Oxidation has reached a slight depth only in these veins and secondary enrichment does not appear to have effected any important rearrangement of the values. The water level in this district is deep, and the slight depth of secondary alteration and oxidation must be explained by the geologic youth of the region, erosion having proceeded more rapidly than oxidation. The degree to which the rocks have been shattered and the presence of abundant druse-lined cavities in the veins, together with the type of rock alteration, indicate a slight depth at the time of ore deposition, and that no great depth of rock has been removed by erosion since the veins were formed. The oxidized ores found at the surface contained chiefly malachite, cerargyrite, and gold, associated with limonite.

[1] L. C. Graton, *P. P.* 68, U. S. G. S., p. 195.

INDEX

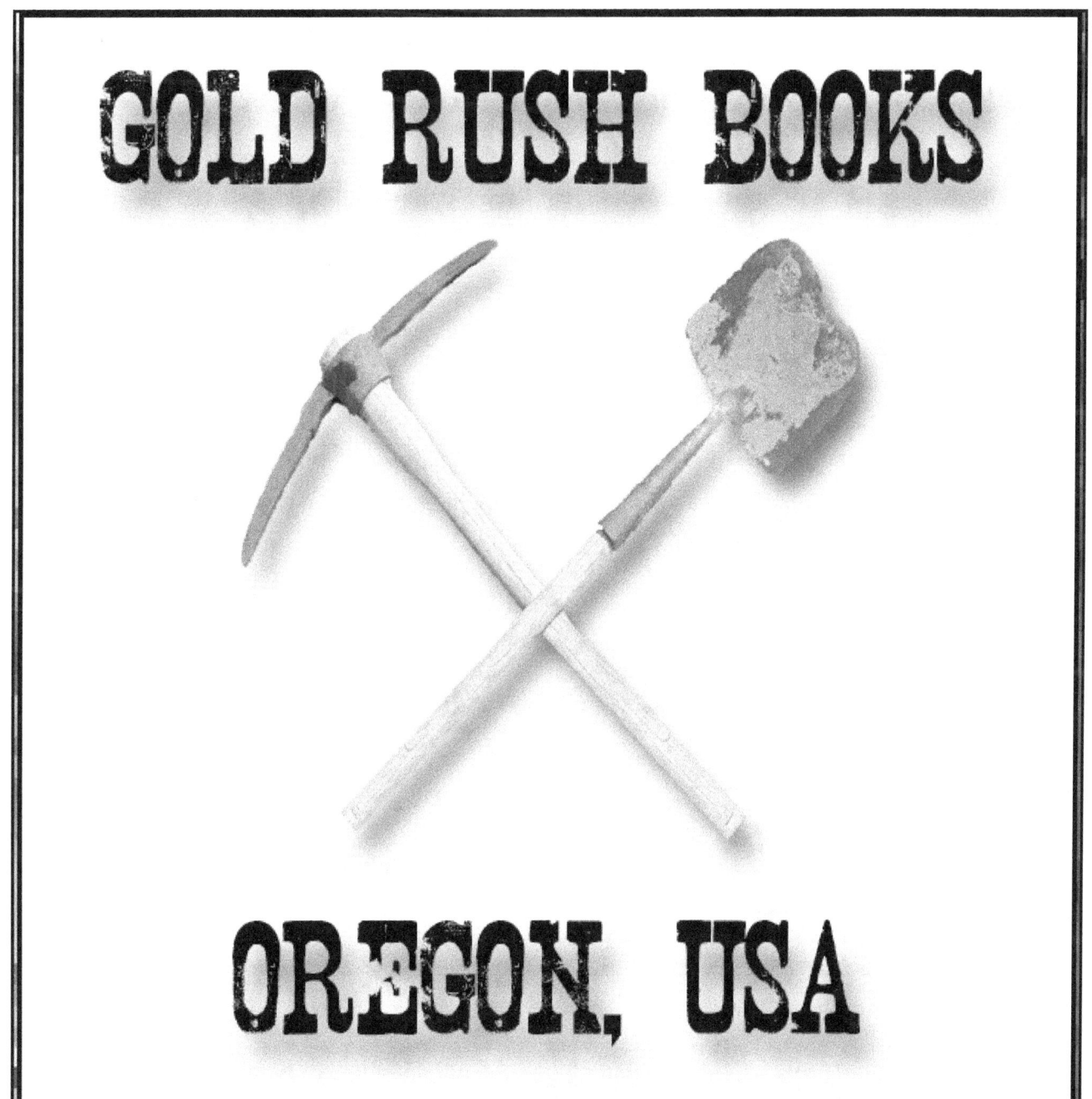

www.GoldMiningBooks.com

Books On Mining

Visit: www.goldminingbooks.com to order your copies or ask your favorite book seller to offer them.

Mining Books by Kerby Jackson

<u>Gold Dust: Stories From Oregon's Mining Years</u> - Oregon mining historian and prospector, Kerby Jackson, brings you a treasure trove of seventeen stories on Southern Oregon's rich history of gold prospecting, the prospectors and their discoveries, and the breathtaking areas they settled in and made homes. 5" X 8", 98 ppgs. **Retail Price: $11.99**

<u>The Golden Trail: More Stories From Oregon's Mining Years</u> - In his follow-up to "Gold Dust: Stories of Oregon's Mining Years", this time around, Jackson brings us twelve tales from Oregon's Gold Rush, including the story about the first gold strike on Canyon Creek in Grant County, about the old timers who found gold by the pail full at the Victor Mine near Galice, how Iradel Bray discovered a rich ledge of gold on the Coquille River during the height of the Rogue River War, a tale of two elderly miners on the hunt for a lost mine in the Cascade Mountains, details about the discovery of the famous Armstrong Nugget and others. 5" X 8", 70 ppgs. **Retail Price: $10.99**

Oregon Mining Books

<u>Geology and Mineral Resources of Josephine County, Oregon</u> - Unavailable since the 1970's, this important publication was originally compiled by the Oregon Department of Geology and Mineral Industries and includes important details on the economic geology and mineral resources of this important mining area in South Western Oregon. Included are notes on the history, geology and development of important mines, as well as insights into the mining of gold, copper, nickel, limestone, chromium and other minerals found in large quantities in Josephine County, Oregon. 8.5" X 11", 54 ppgs. **Retail Price: $9.99**

<u>Mines and Prospects of the Mount Reuben Mining District</u> - Unavailable since 1947, this important publication was originally compiled by geologist Elton Youngberg of the Oregon Department of Geology and Mineral Industries and includes detailed descriptions, histories and the geology of the Mount Reuben Mining District in Josephine County, Oregon. Included are notes on the history, geology, development and assay statistics, as well as underground maps of all the major mines and prospects in the vicinity of this much neglected mining district. 8.5" X 11", 48 ppgs. **Retail Price: $9.99**

<u>The Granite Mining District</u> - Notes on the history, geology and development of important mines in the well known Granite Mining District which is located in Grant County, Oregon. Some of the mines discussed include the Ajax, Blue Ribbon, Buffalo, Continental, Cougar-Independence, Magnolia, New York, Standard and the Tillicum. Also included are many rare maps pertaining to the mines in the area. 8.5" X 11", 48 ppgs. **Retail Price: $9.99**

<u>Ore Deposits of the Takilma and Waldo Mining Districts of Josephine County, Oregon</u> - The Waldo and Takilma mining districts are most notable for the fact that the earliest large scale mining of placer gold and copper in Oregon took place in these two areas. Included are details about some of the earliest large gold mines in the state such as the Llano de Oro, High Gravel, Cameron, Platerica, Deep Gravel and others, as well as copper mines such as the famous Queen of Bronze mine, the Waldo, Lily and Cowboy mines. This volume also includes six maps and 20 original illustrations. 8.5" X 11", 74 ppgs. **Retail Price: $9.99**

<u>Metal Mines of Douglas, Coos and Curry Counties, Oregon</u> - Oregon mining historian Kerby Jackson introduces us to a classic work on Oregon's mining history in this important re-issue of Bulletin 14C Volume 1, otherwise known as the Douglas, Coos & Curry Counties, Oregon Metal Mines Handbook. Unavailable since 1940, this important publication was originally compiled by the Oregon Department of Geology and Mineral Industries includes detailed descriptions, histories and the geology of over 250 metallic mineral mines and prospects in this rugged area of South West Oregon. 8.5" X 11", 158 ppgs. **Retail Price: $19.99**

Metal Mines of Jackson County, Oregon - Unavailable since 1943, this important publication was originally compiled by the Oregon Department of Geology and Mineral Industries includes detailed descriptions, histories and the geology of over 450 metallic mineral mines and prospects in Jackson County, Oregon. Included are such famous gold mining areas as Gold Hill, Jacksonville, Sterling and the Upper Applegate. 8.5" X 11", 220 ppgs. Retail Price: $24.99

Metal Mines of Josephine County, Oregon - Oregon mining historian Kerby Jackson introduces us to a classic work on Oregon's mining history in this important re-issue of Bulletin 14C, otherwise known as the Josephine County, Oregon Metal Mines Handbook. Unavailable since 1952, this important publication was originally compiled by the Oregon Department of Geology and Mineral Industries includes detailed descriptions, histories and the geology of over 500 metallic mineral mines and prospects in Josephine County, Oregon. 8.5" X 11", 250 ppgs. Retail Price: $24.99

Metal Mines of North East Oregon - Oregon mining historian Kerby Jackson introduces us to a classic work on Oregon's mining history in this important re-issue of Bulletin 14A and 14B, otherwise known as the North East Oregon Metal Mines Handbook. Unavailable since 1941, this important publication was originally compiled by the Oregon Department of Geology and Mineral Industries and includes detailed descriptions, histories and the geology of over 750 metallic mineral mines and prospects in North Eastern Oregon. 8.5" X 11", 310 ppgs. Retail Price: $29.99

Metal Mines of North West Oregon - Oregon mining historian Kerby Jackson introduces us to a classic work on Oregon's mining history in this important re-issue of Bulletin 14D, otherwise known as the North West Oregon Metal Mines Handbook. Unavailable since 1951, this important publication was originally compiled by the Oregon Department of Geology and Mineral Industries and includes detailed descriptions, histories and the geology of over 250 metallic mineral mines and prospects in North Western Oregon. 8.5" X 11", 182 ppgs. Retail Price: $19.99

Mines and Prospects of Oregon - Mining historian Kerby Jackson introduces us to a classic mining work by the Oregon Bureau of Mines in this important re-issue of The Handbook of Mines and Prospects of Oregon. Unavailable since 1916, this publication includes important insights into hundreds of gold, silver, copper, coal, limestone and other mines that operated in the State of Oregon around the turn of the 19th Century. Included are not only geological details on early mines throughout Oregon, but also insights into their history, production, locations and in some cases, also included are rare maps of their underground workings. 8.5" X 11", 314 ppgs. Retail Price: $24.99

Lode Gold of the Klamath Mountains of Northern California and South West Oregon
(See California Mining Books)

Mineral Resources of South West Oregon - Unavailable since 1914, this publication includes important insights into dozens of mines that once operated in South West Oregon, including the famous gold fields of Josephine and Jackson Counties, as well as the Coal Mines of Coos County. Included are not only geological details on early mines throughout South West Oregon, but also insights into their history, production and locations. 8.5" X 11", 154 ppgs. Retail Price: $11.99

Chromite Mining in The Klamath Mountains of California and Oregon
(See California Mining Books)

Southern Oregon Mineral Wealth - Unavailable since 1904, this rare publication provides a unique snapshot into the mines that were operating in the area at the time. Included are not only geological details on early mines throughout South West Oregon, but also insights into their history, production and locations. Some of the mining areas include Grave Creek, Greenback, Wolf Creek, Jump Off Joe Creek, Granite Hill, Galice, Mount Reuben, Gold Hill, Galls Creek, Kane Creek, Sardine Creek, Birdseye Creek, Evans Creek, Foots Creek, Jacksonville, Ashland, the Applegate River, Waldo, Kerby and the Illinois River, Althouse and Sucker Creek, as well as insights into local copper mining and other topics. 8.5" X 11", 64 ppgs. Retail Price: $8.99

Geology and Ore Deposits of the Takilma and Waldo Mining Districts - Unavailable since the 1933, this publication was originally compiled by the United States Geological Survey and includes details on gold and copper mining in the Takilma and Waldo Districts of Josephine County, Oregon. The Waldo and Takilma mining districts are most notable for the fact that the earliest large scale mining of placer gold and copper in Oregon took place in these two areas. Included in this report are details about some of the earliest large gold mines in the state such as the Llano de Oro, High Gravel, Cameron, Platerica, Deep Gravel and others, as well as copper mines such as the famous Queen of Bronze mine, the Waldo, Lily and Cowboy mines. In addition to geological examinations, insights are also provided into the production, day to day operations and early histories of these mines, as well as calculations of known mineral reserves in the area. This volume also includes six maps and 20 original illustrations. 8.5" X 11", 74 ppgs. Retail Price: $9.99

Gold Mines of Oregon - Oregon mining historian Kerby Jackson introduces us to a classic work on Oregon's mining history in this important re-issue of Bulletin 61, otherwise known as "Gold and Silver In Oregon". Unavailable since 1968, this important publication was originally compiled by geologists Howard C. Brooks and Len Ramp of the Oregon Department of Geology and Mineral Industries and includes detailed descriptions, histories and the geology of over 450 gold mines Oregon. Included are notes on the history, geology and gold production statistics of all the major mining areas in Oregon including the Klamath Mountains, the Blue Mountains and the North Cascades. While gold is where you find it, as every miner knows, the path to success is to prospect for gold where it was previously found. 8.5" X 11", 344 ppgs. Retail Price: $24.99

Mines and Mineral Resources of Curry County Oregon - Originally published in 1916, this important publication on Oregon Mining has not been available for nearly a century. Included are rare insights into the history, production and locations of dozens of gold mines in Curry County, Oregon, as well as detailed information on important Oregon mining districts in that area such as those at Agness, Bald Face Creek, Mule Creek, Boulder Creek, China Diggings, Collier Creek, Elk River, Gold Beach, Rock Creek, Sixes River and elsewhere. Particular attention is especially paid to the famous beach gold deposits of this portion of the Oregon Coast. 8.5" X 11", 140 ppgs. Retail Price: $11.99

Chromite Mining in South West Oregon - Originally published in 1961, this important publication on Oregon Mining has not been available for nearly a century. Included are rare insights into the history, production and locations of nearly 300 chromite mines in South Western Oregon. 8.5" X 11", 184 ppgs. Retail Price: $14.99

Mineral Resources of Douglas County Oregon - Originally published in 1972, this important publication on Oregon Mining has not been available for nearly forty years. Included are rare insights into the geology, history, production and locations of numerous gold mines and other mining properties in Douglas County, Oregon. 8.5" X 11", 124 ppgs. Retail Price: $11.99

Mineral Resources of Coos County Oregon - Originally published in 1972, this important publication on Oregon Mining has not been available for nearly forty years. Included are rare insights into the geology, history, production and locations of numerous gold mines and other mining properties in Coos County, Oregon. 8.5" X 11", 100 ppgs. Retail Price: $11.99

Mineral Resources of Lane County Oregon - Originally published in 1938, this important publication on Oregon Mining has not been available for nearly seventy five years. Included are extremely rare insights into the geology and mines of Lane County, Oregon, in particular in the Bohemia, Blue River, Oakridge, Black Butte and Winberry Mining Districts. 8.5" X 11", 82 ppgs. Retail Price: $9.99

Mineral Resources of the Upper Chetco River of Oregon: Including the Kalmiopsis Wilderness - Originally published in 1975, this important publication on Oregon Mining has not been available for nearly forty years. Withdrawn under the 1872 Mining Act since 1984, real insight into the minerals resources and mines of the Upper Chetco River has long been unavailable due to the remoteness of the area. Despite this, the decades of battle between property owners and environmental extremists over the last private mining inholding in the area has continued to pique the interest of those interested in mining and other forms of natural resource use. Gold mining began in the area in the 1850's and has a rich history in this geographic area, even if the facts surrounding it are little known. Included are twenty two rare photographs, as well as insights into the Becca and Morning Mine, the Emmly Mine (also known as Emily Camp), the Frazier Mine, the Golden Dream or Higgins Mine, Hustis Mine, Peck Mine and others. 8.5" X 11", 64 ppgs. Retail Price: $8.99

Gold Dredging in Oregon - Originally published in 1939, this important publication on Oregon Mining has not been available for nearly seventy five years. Included are extremely rare insights into the history and day to day operations of the dragline and bucketline gold dredges that once worked the placer gold fields of South West and North East Oregon in decades gone by. Also included are details into the areas that were worked by gold dredges in Josephine, Jackson, Baker and Grant counties, as well as the economic factors that impacted this mining method. This volume also offers a unique look into the values of river bottom land in relation to both farming and mining, in how farm lands were mined, re-soiled and reclamated after the dredges worked them. Featured are hard to find maps of the gold dredge fields, as well as rare photographs from a bygone era. 8.5" X 11", 86 ppgs. Retail Price: $8.99

Quick Silver Mining in Oregon - Originally published in 1963, this important publication on Oregon Mining has not been available for over fifty years. This publication includes details into the history and production of Elemental Mercury or Quicksilver in the State of Oregon. 8.5" X 11", 238 ppgs. Retail Price: $15.99

Mines of the Greenhorn Mining District of Grant County Oregon - Originally published in 1948, this important publication on Oregon Mining has not been available for over sixty five years. In this publication are rare insights into the mines of the famous Greenhorn Mining District of Grant County, Oregon, especially the famous Morning Mine. Also included are details on the Tempest, Tiger, Bi-Metallic, Windsor, Psyche, Big Johnny, Snow Creek, Banzette and Paramount Mines, as well as prospects in the vicinities in the famous mining areas of Mormon Basin, Vinegar Basin and Desolation Creek. Included are hard to find mine maps and dozens of rare photographs from the bygone era of Grant County's rich mining history. 8.5" X 11", 72 ppgs. Retail Price: $9.99

Geology of the Wallowa Mountains of Oregon: Part I (Volume 1) - Originally published in 1938, this important publication on Oregon Mining has not been available for nearly seventy five years. Included are details on the geology of this unique portion of North Eastern Oregon. This is the first part of a two book series on the area. Accompanying the text are rare photographs and historic maps.8.5" X 11", 92 ppgs. Retail Price: $9.99

Geology of the Wallowa Mountains of Oregon: Part II (Volume 2) - Originally published in 1938, this important publication on Oregon Mining has not been available for nearly seventy five years. Included are details on the geology of this unique portion of North Eastern Oregon. This is the first part of a two book series on the area. Accompanying the text are rare photographs and historic maps.8.5" X 11", 94 ppgs. Retail Price: $9.99

Field Identification of Minerals For Oregon Prospectors - Originally published in 1940, this important publication on Oregon Mining has not been available for nearly seventy five years. Included in this volume is an easy system for testing and identifying a wide range of minerals that might be found by prospectors, geologists and rockhounds in the State of Oregon, as well as in other locales. Topics include how to put together your own field testing kit and how to conduct rudimentary tests in the field. This volume is written in a clear and concise way to make it useful even for beginners. **8.5" X 11", 158 ppgs. Retail Price: $14.99**

The Bohemia Mining District of Oregon - Originally published in 1900, this important publication on Oregon Mining has not been available for over a century. Included in this volume are important insights into the famous Bohemia Mining District of Oregon, including the histories and locations of important gold mines in the area such as the Ophir Mine, Clarence, Acturas, Peek-a-boo, White Swan, Combination Mine, the Musick Mine, The California, White Ghost, The Mystery, Wall Street, Vesuvius, Story, Lizzie Bullock, Delta, Elsie Dora, Golden Slipper, Broadway, Champion Mine, Knott, Noonday, Helena, White Wings, Riverside and others. Also included are notes on the nearby Blue River Mining District. **8.5" X 11", 58 ppgs. Retail Price: $9.99**

The Gold Fields of Eastern Oregon - Unavailable since 1900, this publication was originally compiled by the Baker City Chamber of Commerce Offering important insights into the gold mining history of Eastern Oregon, "The Gold Fields of Eastern Oregon" sheds a rare light on many of the gold mines that were operating at the turn of the 19th Century in Baker County and Grant County in North Eastern Oregon. Some of the areas featured include the Cable Cove District, Baisely-Elhorn, Granite, Red Boy, Bonanza, Susanville, Sparta, Virtue, Vaughn, Sumpter, Burnt River, Rye Valley and other mining districts. Included is basic information on not only many gold mines that are well known to those interested in Eastern Oregon mining history, but also many mines and prospects which have been mostly lost to the passage of time. Accompanying are numerous rare photos 8.5" X 11", 78 ppgs. Retail Price: $10.99

Gold Mining in Eastern Oregon - Originally published in 1938, this important publication on Oregon Mining has not been available for over a century. Included in this volume are important insights into the famous mining districts of Eastern Oregon during the late 1930's. Particular attention is given to those gold mines with milling and concentrating facilities in the Greenhorn, Red Boy, Alamo, Bonanza, Granite, Cable Cove, Cracker Creek, Virtue, Keating, Medical Springs, Sanger, Sparta, Chicken Creek, Mormon Basin, Connor Creek, Cornucopia and the Bull Run Mining Districts. Some of the mines featured include the Ben Harrison, North Pole-Columbia, Highland Maxwell, Baisley-Elkhorn, White Swan, Balm Creek, Twin Baby, Gem of Sparta, New Deal, Gleason, Gifford-Johnson, Cornucopia, Record, Bull Run, Orion and others. Of particular interest are the mill flow sheets and descriptions of milling operations of these mines. 8.5" X 11", 68 ppgs. Retail Price: $8.99

The Gold Belt of the Blue Mountains of Oregon - Originally published in 1901, this important publication on Oregon Mining has not been available for over a century. Included in this volume are rare insights into the gold deposits of the Blue Mountains of North East Oregon, including the history of their early discovery and early production. Extensive details are offered on this important mining area's mineralogy and economic geology, as well as insights into nearby gold placers, silver deposits and copper deposits. Featured are the Elkhorn and Rock Creek mining districts, the Pocahontas district, Auburn and Minersville districts, Sumpter and Cracker Creek, Cable Cove, the Camp Carson district, Granite, Alamo, Greenhorn, Robinsonville, the Upper Burnt River Valley and Bonanza districts, Susanville, Quartzburg, Canyon Creek, Virtue, the Copper Butte district, the North Powder River, Sparta, Eagle Creek, Cornucopia, Pine Creek, Lower Powder River, the Upper Snake River Canyon, Rye Valley, Lower Burnt River Valley, Mormon Basin, the Malheur and Clarks Creek districts, Sutton Creek and others. Of particular interest are important details on numerous gold mines and prospects in these mining districts, including their locations, histories, geology and other important information, as well as information on silver, copper and fire opal deposits. 8.5" X 11", 250 ppgs. Retail Price: $24.99

Mining in the Cascades Range of Oregon - Originally published in 1938, this important publication on Oregon Mining has not been available for over seventy five years. Included in this volume are rare insights into the gold mines and other types of metal mines in the Cascades Mountain Range of Oregon. Some of the important mining areas covered include the famous Bohemia Mining District, the North Santiam Mining District, Quartzville Mining District, Blue River Mining District, Fall Creek Mining District, Oakridge District, Zinc District, Buzzard-Al Sarena District, Grand Cove, Climax District and Barron Mining District. Of particular interest are important details on over 100 mines and prospects in these mining districts, including their locations, histories, geology and other important information. **8.5" X 11", 170 ppgs. Retail Price: $14.99**

Beach Gold Placers of the Oregon Coast - Originally published in 1934, this important publication on Oregon Mining has not been available for over 80 years. Included in this volume are rare insights into the beach gold deposits of the State of Oregon, including their locations, occurance, composition and geology. Of particular interest is information on placer platinum in Oregon's rich beach deposits. Also included are the locations and other information on some famous Oregon beach mines, including the Pioneer, Eagle, Chickamin, Iowa and beach placer mines north of the mouth of the Rogue River. **8.5" X 11", 60 ppgs. Retail Price: $8.99**

Idaho Mining Books

Gold in Idaho - Unavailable since the 1940's, this publication was originally compiled by the Idaho Bureau of Mines and includes details on gold mining in Idaho. Included is not only raw data on gold production in Idaho, but also valuable insight into where gold may be found in Idaho, as well as practical information on the gold bearing rocks and other geological features that will assist those looking for placer and lode gold in the State of Idaho. This volume also includes thirteen gold maps that greatly enhance the practical usability of the information contained in this small book detailing where to find gold in Idaho. **8.5" X 11", 72 ppgs. Retail Price: $9.99**

Geology of the Couer D'Alene Mining District of Idaho - Unavailable since 1961, this publication was originally compiled by the Idaho Bureau of Mines and Geology and includes details on the mining of gold, silver and other minerals in the famous Coeur D'Alene Mining District in Northern Idaho. Included are details on the early history of the Coeur D'Alene Mining District, local tectonic settings, ore deposit features, information on the mineral belts of the Osburn Fault, as well as detailed information on the famous Bunker Hill Mine, the Dayrock Mine, Galena Mine, Lucky Friday Mine and the infamous Sunshine Mine. This volume also includes sixteen hard to find maps. **8.5" X 11", 70 ppgs. Retail Price: $9.99**

The Gold Camps and Silver Cities of Idaho - Originally published in 1963, this important publication on Idaho Mining has not been available for nearly fifty years. Included are rare insights into the history of Idaho's Gold Rush, as well as the mad craze for silver in the Idaho Panhandle. Documented in fine detail are the early mining excitements at Boise Basin, at South Boise, in the Owyhees, at Deadwood, Long Valley, Stanley Basin and Robinson Bar, at Atlanta, on the famous Boise River, Volcano, Little Smokey, Banner, Boise Ridge, Hailey, Leesburg, Lemhi, Pearl, at South Mountain, Shoup and Ulysses, Yellow Jacket and Loon Creek. The story follows with the appearance of Chinese miners at the new mining camps on the Snake River, Black Pine, Yankee Fork, Bay Horse, Clayton, Heath, Seven Devils, Gibbonsville, Vienna and Sawtooth City. Also included are special sections on the Idaho Lead and Silver mines of the late 1800's, as well as the mining discoveries of the early 1900's that paved the way for Idaho's modern mining and mineral industry. Lavishly illustrated with rare historic photos, this volume provides a one of a kind documentary into Idaho's mining history that is sure to be enjoyed by not only modern miners and prospectors who still scour the hills in search of nature's treasures, but also those enjoy history and tromping through overgrown ghost towns and long abandoned mining camps. **8.5" X 11", 186 ppgs. Retail Price: $14.99**

Ore Deposits and Mining in North Western Custer County Idaho - Unavailable since 1913, this important publication was originally published by the Us Department of the Interior and has been unavailable for a century. Included are fine details on the geology, geography, gold placers and gold and silver bearing quartz veins of the mining region of North West Custer County, Idaho. Of particular interest is a rare look at the mines and prospects of the region, including those such as the Ramshorn Mine, SkyLark, Riverview, Excelsior, Beardsley, Pacific, Hoosier, Silver Brick, Forest Rose and dozens of others in the Bay Horse Mining District. Also covered are the mines of the Yankee Fork District such as the Lucky Boy, Badger, Black, Enterprise, Charles Dickens, Morrison, Golden Sunbeam, Montana, Golden Gate and others, as well as those in the Loon Mining District. **8.5" X 11", 126 ppgs. Retail Price: $12.99**

Gold Rush To Idaho - Unavailable since 1963, this important publication was originally published by the Idaho Bureau of Mines and has been unavailable for 50 years. "Gold Rush To Idaho" revisits the earliest years of the discovery of gold in Idaho Territory and introduces us to the conditions that the pioneer gold seekers met when they blazed a trail through the wilderness of Idaho's mountains and discovered the precious yellow metal at Oro Fino and Pierce. Subsequent rushes followed at places like Elk City, Newsome, Clearwater Station, Florence, Warrens and elsewhere. Of particular interest is a rare look at the hardships that the first miners in Idaho met with during their day to day existences and their attempts to bring law and order to their mining camps. 8.5" X 11", 88 ppgs. Retail Price: $9.99

The Geology and Mines of Northern Idaho and North Western Montana - Unavailable since 1909, this important publication was originally published by the Us Department of the Interior and has been unavailable for a century. Included are fine details on the geology and geography of the mining regions of Northern Idaho and North Western Montana. Of particular interest is a rare look at the mines and prospects of the region, including those in the Pine Creek Mining District, Lake Pend Oreille district, Troy Mining District, Sylvanite District, Cabinet Mining District, Prospect Mining District and the Missoula Valley. Some of the mines featured include the Iron Mountain, Silver Butte, Snowshoe, Grouse Mountain Mine and others. 8.5" X 11", 142 ppgs. Retail Price: $12.99

Mining in the Alturas Quadrangle of Blaine County Idaho - Unavailable since 1922, this important publication was originally published by the Idaho Bureau of Mines and has been unavailable for ninety years. Topics include the geology, rock formations and the formation of ore deposits in this important mining area of Idaho. Of particular focus is information on the local geology, quartz veins and ore deposits of this portion of Idaho. Included are hard to find details, including the descriptions and locations of numerous gold and silver mines in the area including the Silver King, Pilgrim, Columbia, Lone Jack, Sunbeam, Pride of the West, Lucky Boy, Scotia, Atlanta, Beaver-Bidwell and others mines and prospects. 8.5" X 11", 56 ppgs. Retail Price: $8.99

Mining in Lemhi County Idaho - Originally published in 1913, this important book on Idaho Mining has not been available to miners for over a century. Included are rare insights into hundreds of gold, silver, copper and other mines in this famous Idaho mining area. Details include the locations, geology, history, production and other facts of the mines of this region, not only gold and silver hardrock mines, but also gold placer mines, lead-silver deposits, copper mines, cobalt-nickel deposits, tungsten and tin mines . It is lavishly illustrated with hard to find photos of the period and rare mining maps. Some of the vicinities featured include the Nicholia Mining District, Spring Mountain District, Texas District, Blue Wing District, Junction District, McDevitt District, Pratt Creek, Eldorado District, Kirtley Creek, Carmen Creek, Gibbonsville, Indian Creek, Mineral Hill District, Mackinaw, Eureka District, Blackbird District, YellowJacket District, Gravel Range District, Junction District, Parker Mountain and other mining districts. 8.5" X 11", 226 ppgs. Retail Price: $19.99

Utah Mining Books

Fluorite in Utah - Unavailable since 1954, this publication was originally compiled by the USGS, State of Utah and U.S. Atomic Energy Commission and details the mining of fluorspar, also known as fluorite in the State of Utah. Included are details on the geology and history of fluorspar (fluorite) mining in Utah, including details on where this unique gem mineral may be found in the State of Utah. 8.5" X 11", 60 ppgs. Retail Price: $8.99

California Mining Books

The Tertiary Gravels of the Sierra Nevada of California - Mining historian Kerby Jackson introduces us to a classic mining work by Waldemar Lindgren in this important re-issue of The Tertiary Gravels of the Sierra Nevada of California. Unavailable since 1911, this publication includes details on the gold bearing ancient river channels of the famous Sierra Nevada region of California. 8.5" X 11", 282 ppgs. Retail Price: $19.99

The Mother Lode Mining Region of California - Unavailable since 1900, this publication includes details on the gold mines of California's famous Mother Lode gold mining area. Included are details on the geology, history and important gold mines of the region, as well as insights into historic mining methods, mine timbering, mining machinery, mining bell signals and other details on how these mines operated. Also included are insights into the gold mines of the California Mother Lode that were in operation during the first sixty years of California's mining history. 8.5" X 11", 176 ppgs. Retail Price: $14.99

Lode Gold of the Klamath Mountains of Northern California and South West Oregon - Unavailable since 1971, this publication was originally compiled by Preston E. Hotz and includes details on the lode mining districts of Oregon and California's Klamath Mountains. Included are details on the geology, history and important lode mines of the French Gulch, Deadwood, Whiskeytown, Shasta, Redding, Muletown, South Fork, Old Diggings, Dog Creek (Delta), Bully Choop (Indian Creek), Harrison Gulch, Hayfork, Minersville, Trinity Center, Canyon Creek, East Fork, New River, Denny, Liberty (Black Bear), Cecilville, Callahan, Yreka, Fort Jones and Happy Camp mining districts in California, as well as the Ashland, Rogue River, Applegate, Illinois River, Takilma, Greenback, Galice, Silver Peak, Myrtle Creek and Mule Creek districts of South Western Oregon. Also included are insights into the mineralization and other characteristics of this important mining region. 8.5" X 11", 100 ppgs. Retail Price: $10.99

Mines and Mineral Resources of Shasta County, Siskiyou County, Trinity County: California - Unavailable since 1915, this publication was originally compiled by the California State Mining Bureau and includes details on the gold mines of this area of Northern California. Also included are insights into the mineralization and other characteristics of this important mining region, as well as the location of historic gold mines. **8.5" X 11", 204 ppgs. Retail Price: $19.99**

Geology of the Yreka Quadrangle, Siskiyou County, California - Unavailable since 1977, this publication was originally compiled by Preston E. Hotz and includes details on the geology of the Yreka Quadrangle of Siskiyou County, California. Also included are insights into the mineralization and other characteristics of this important mining region. **8.5" X 11", 78 ppgs. Retail Price: $7.99**

Mines of San Diego and Imperial Counties, California - Originally published in 1914, this important publication on California Mining has not been available for a century. This publication includes important information on the early gold mines of San Diego and Imperial County, which were some of the first gold fields mined in California by early Spanish and Mexican miners before the 49ers came on the scene. Included are not only details on early mining methods in the area, production statistics and geological information, but also the location of the early gold mines that helped make California "The Golden State". Also included are details on the mining of other minerals such as silver, lead, zinc, manganese, tungsten, vanadium, asbestos, barite, borax, cement, clay, dolomite, fluospar, gem stones, graphite, marble, salines, petroleum, stronium, talc and others. **8.5" X 11", 116 ppgs. Retail Price: $12.99**

Mines of Sierra County, California - Unavailable since 1920, this publication was originally compiled by the California State Mining Bureau and includes details on the gold mines of Sierra County, California. Also included are insights into the mineralization and other characteristics of this important mining region, as well as the location of historic gold mines. **8.5" X 11", 156 ppgs. Retail Price: $19.99**

Mines of Plumas County, California - Unavailable since 1918, this publication was originally compiled by the California State Mining Bureau and includes details on the gold mines of Plumas County, California. Also included are insights into the mineralization and other characteristics of this important mining region, as well as the location of historic gold mines. **8.5" X 11", 200 ppgs. Retail Price: $19.99**

Mines of El Dorado, Placer, Sacramento and Yuba Counties, California - Originally published in 1917, this important publication on California Mining has not been available for nearly a century. This publication includes important information on the early gold mines of El Dorado County, Placer County, Sacramento County and Yuba County, which were some of the first gold fields mined by the Forty-Niners during the California Gold Rush. Included are not only details on early mining methods in the area, production statistics and geological information, but also the location of the early gold mines that helped make California "The Golden State". Also included are insights into the early mining of chrome, copper and other minerals in this important mining area. **8.5" X 11", 204 ppgs. Retail Price: $19.99**

Mines of Los Angeles, Orange and Riverside Counties, California - Originally published in 1917, this important publication on California Mining has not been available for nearly a century. This publication includes important information on the early gold mines of Los Angeles County, Orange County and Riverside County, which were some of the first gold fields mined in California by early Spanish and Mexican miners before the 49ers came on the scene. Included are not only details on early mining methods in the area, production statistics and geological information, but also the location of the early gold mines that helped make California "The Golden State". **8.5" X 11", 146 ppgs. Retail Price: $12.99**

Mines of San Bernadino and Tulare Counties, California - Originally published in 1917, this important publication on California Mining has not been available for nearly a century. This publication includes important information on the early gold mines of San Bernadino and Tulare County, which were some of the first gold fields mined in California by early Spanish and Mexican miners before the 49ers came on the scene. Included are not only details on early mining methods in the area, production statistics and geological information, but also the location of the early gold mines that helped make California "The Golden State". Also included are details on the mining of other minerals such as copper, iron, lead, zinc, manganese, tungsten, vanadium, asbestos, barite, borax, cement, clay, dolomite, fluospar, gem stones, graphite, marble, salines, petroleum, stronium, talc and others. **8.5" X 11", 200 ppgs. Retail Price: $19.99**

Chromite Mining in The Klamath Mountains of California and Oregon - Unavailable since 1919, this publication was originally compiled by J.S. Diller of the United States Department of Geological Survey and includes details on the chromite mines of this area of Northern California and Southern Oregon. Also included are insights into the mineralization and other characteristics of this important mining region, as well as the location of historic mines. Also included are insights into chromite mining in Eastern Oregon and Montana. **8.5" X 11", 98 ppgs. Retail Price: $9.99**

Mines and Mining in Amador, Calaveras and Tuolumne Counties, California - Unavailable since 1915, this publication was originally compiled by William Tucker and includes details on the mines and mineral resources of this important California mining area. Included are details on the geology, history and important gold mines of the region, as well as insights into other local mineral resources such as asbestos, clay, copper, talc, limestone and others. Also included are insights into the mineralization and other characteristics of this important portion of California's Mother Lode mining region. 8.5" X 11", 198 ppgs. Retail Price: $14.99

The Cerro Gordo Mining District of Inyo County California - Unavailable since 1963, this publication was originally compiled by the United States Department of Interior. Included are insights into the mineralization and other characteristics of this important mining region of Southern California. Topics include the mining of gold and silver in this important mining district in Inyo County, California, including details on the history, production and locations of the Cerro Gordo Mine, the Morning Star Mine, Estelle Tunnel, Charles Lease Tunnel, Ignacio, Hart, Crosscut Tunnel, Sunset, Upper Newtown, Newtown, Ella, Perseverance, Newsboy, Belmont and other silver and gold mines in the Cerro Gordo Mining District. This volume also includes important insights into the fossil record, geologic formations, faults and other aspects of economic geology in this California mining district. 8.5" X 11", 104 ppgs. Retail Price: $10.99

Mining in Butte, Lassen, Modoc, Sutter and Tehama Counties of California - Unavailable since 1917, this publication was originally compiled by the United States Department of Interior. Included are insights into the mineralization and other characteristics of this important mining region of California. Topics include the mining of asbestos, chromite, gold, diamonds and manganese in Butte County, the mining of gold and copper in the Hayden Hill and Diamond Mountain mining districts of Lassen County, the mining of coal, salt, copper and gold in the High Grade and Winters mining districts of Modoc County, gold mining in Sutter County and the mining of gold, chromite, manganese and copper in Tehama County. This volume also includes the production records and locations of numerous mines in this important mining region. 8.5" X 11", 114 ppgs. Retail Price: $11.99

Mines of Trinity County California - Originally published in 1965, this important publication on California Mining has not been available for nearly fifty years. This publication includes important information on mines and mining in Trinity County, California, as well insights into the mineralization and geology of this important mining area in Northern California. Included are extensive details on hardrock and placer gold mines and prospects, including charts showing the locations of these historic mines.. 8.5" X 11", 144 ppgs. Retail Price: $12.99

Mines of Kern County California - Originally published in 1962, this important publication on California Mining has not been available for nearly fifty years. This publication includes important information on mines and mining in Kern County, California, as well insights into the mineralization and geology of this important mining area in California. Included are extensive details on hardrock and placer gold mines and prospects, including charts showing the locations of these historic mines. 8.5" X 11", 398 ppgs. Retail Price: $24.99

Mines of Calaveras County California - Originally published in 1962, this important publication on California Mining has not been available for nearly fifty years. This publication includes important information on mines and mining in Calaveras County, California, as well insights into the mineralization and geology of this important mining area in Northern California. Included are extensive details on hardrock and placer gold mines and prospects, including charts showing the locations of these historic mines. 8.5" X 11", 236 ppgs. Retail Price: $19.99

Lode Gold Mining in Grass Valley California - Unavailable since 1940, this publication was originally compiled by the United States Department of Interior. Included are insights into the gold mineralization and other characteristics of this important mining region of Nevada County, California. This volume also includes important insights into the geologic formations, faults and other aspects of economic geology in this California mining district. Of particular interest are the fine details on many hardrock gold mines in the area, including their locations, histories, development and mineralization. Some of the mines featured include the Gold Hill Mine, Massachusetts Hill, Boundary, Peabody, Golden Center, North Star, Omaha, Lone Jack, Homeward Bound, Hartery, Wisconsin, Allison Ranch, Phoenix, Kate Hayes, W.Y.O.D., Empire, Rich Hill, Daisy Hill, Orleans, Sultana, Centennial, Conlin, Ben Franklin, Crown Point and many others. 8.5" X 11", 148 ppgs. Retail Price: $12.99

Lode Mining in the Alleghany District of Sierra County California - Unavailable since 1913, this publication was originally compiled by the United States Department of Interior. Included are insights into the mineralization and other characteristics of this important mining region of Sierra County. Included are details on the history, production and locations of numerous hardrock gold mines in this famous California area, including the Tightner Mine, Minnie D., Osceola, Eldorado, Twenty One, Sherman, Kenton, Oriental, Rainbow, Plumbago, Irelan, Gold Canyon, North Fork, Federal, Kate Hardy and others. This volume also includes important insights into the fossil record, geologic formations, faults and other aspects of economic geology in this California mining district. 8.5" X 11", 48 ppgs. Retail Price: $7.99

Six Months In The Gold Mines During The California Gold Rush - Unavailable since 1850, this important work is a first hand account of one "49'ers" personal experience during the great California Gold Rush, shedding important light on one of the most exciting periods in the history of not only California, but also the world. Compiled from journals written between 1847 and 1849 by E. Gould Buffum, a native of New York, "Six Months In The Gold Mines During The California Gold Rush" offers a rare look into the day to day lives of the people who came to California to work in her gold mines when the state was still a great frontier. 8.5" X 11", 290 ppgs. **Retail Price: $19.99**

Quartz Mines of the Grass Valley Mining District of California - Unavailable since 1867, this important publication has not been available since those days. This rare publication offers a short dissertation on the early hardrock mines in this important mining district in the California Mother Lode region between the 1850's and 1860's. Also included are hard to find details on the mineralization and locations of these mines, as well as how they were operated in those day. 8.5" X 11", 44 ppgs. **Retail Price: $8.99**

Alaska Mining Books

Ore Deposits of the Willow Creek Mining District, Alaska - Unavailable since 1954, this hard to find publication includes valuable insights into the Willow Creek Mining District near Hatcher Pass in Alaska. The publication includes insights into the history, geology and locations of the well known mines in the area, including the Gold Cord, Independence, Fern, Mabel, Lonesome, Snowbird, Schroff-O'Neil, High Grade, Marion Twin, Thorpe, Webfoot, Kelly-Willow, Lane, Holland and others. 8.5" X 11", 96 ppgs. **Retail Price: $9.99**

The Juneau Gold Belt of Alaska - Unavailable since 1906, this hard to find publication includes valuable insights into the gold mines around Juneau, Alaska. The publication includes important details into the history, geology and locations of the well known gold mines and prospects in the area, including those around Windham Bay, Holkham Bay, Port Snettisham, on Grindstone and Rhine Creeks, Gold Creek, Douglas Island, Salmon Creek, Lemon Creek, Nugget Creek, from the Mendenhall River to Berners Bay, McGinnis Creek, Montana Creek, Peterson Creek, Windfall Creek, the Eagle River, Yankee Basin, Yankee Curve, Kowee Creek and elsewhere. Not only are gold placer mines included, but also hardrock gold mines. 8.5" X 11", 224 ppgs. **Retail Price: $19.99**

Arizona Mining Books

Mines and Mining in Northern Yuma County Arizona - Originally published in 1911, this important publication on Arizona Mining has not been available for over a hundred years. Included are rare insights into the gold, silver, copper and quicksilver mines of Yuma County, Arizona together with hard to find maps and photographs. Some of the mines and mining districts featured include the Planet Copper Mine, Mineral Hill, the Clara Consolidated Mine, Viati Mine, Copper Basin prospect, Bowman Mine, Quartz King, Billy Mack, Carnation, the Wardwell and Osbourne, Valensuella Copper, the Mariquita, Colonial Mine, the French American, the New York-Plomosa, Guadalupe, Lead Camp, Mudersbach Copper Camp, Yellow Bird, the Arizona Northern (Salome Strike), Bonanza (Harqua Hala), Golden Eagle, Hercules, Socorro and others. 8.5" X 11", 144 ppgs. **Retail Price: $11.99**

The Aravaipa and Stanley Mining Districts of Graham County Arizona - Originally published in 1925, this important publication on Arizona Mining has not been available for nearly ninety years. Included are rare insights into the gold and silver mines of these two important mining districts, together with hard to find maps. 8.5" X 11", 140 ppgs. **Retail Price: $11.99**

Gold in the Gold Basin and Lost Basin Mining Districts of Mohave County, Arizona - This volume contains rare insights into the geology and gold mineralization of the Gold Basin and Lost Basin Mining Districts of Mohave County, Arizona that will be of benefit to miners and prospectors. Also included is a significant body of information on the gold mines and prospects of this portion of Arizona. This volume is lavishly illustrated with rare photos and mining maps. 8.5" X 11", 188 ppgs. **Retail Price: $19.99**

Mines of the Jerome and Bradshaw Mountains of Arizona - This important publication on Arizona Mining has not been available for ninety years. This volume contains rare insights into the geology and ore deposits of the Jerome and Bradshaw Mountains of Arizona that will be of benefit to miners and prospectors who work those areas. Included is a significant body of information on the mines and prospects of the Verde, Black Hills, Cherry Creek, Prescott, Walker, Groom Creek, Hassayampa, Bigbug, Turkey Creek, Agua Fria, Black Canyon, Peck, Tiger, Pine Grove, Bradshaw, Tintop, Humbug and Castle Creek Mining Districts. This volume is lavishly illustrated with rare photos and mining maps. 8.5" X 11", 218 ppgs. **Retail Price: $19.99**

The Ajo Mining District of Pima County Arizona - This important publication on Arizona Mining has not been available for nearly seventy years. This volume contains rare insights into the geology and mineralization of the Ajo Mining District in Pima County, Arizona and in particular the famous New Cornelia Mine. 8.5" X 11", 126 ppgs. **Retail Price: $11.99**

<u>Mining in the Santa Rita and Patagonia Mountains of Arizona</u> - Originally published in 1915, this important publication on Arizona Mining has not been available for nearly a century. Included are rare insights into hundreds of gold, silver, copper and other mines in this famous Arizona mining area. Details include the locations, geology, history, production and other facts of the mines of this region. **8.5" X 11", 394 ppgs. Retail Price: $24.99**

<u>Mining in the Bisbee Quadrangle of Arizona</u> - Originally published in 1906, this important publication on Arizona Mining has not been available for nearly a century. Included are rare insights into hundreds of gold, silver, copper and other mines in this famous Arizona mining area. Details include the locations, geology, history, production and other facts of the mines of this important mining region. **8.5" X 11", 188 ppgs. Retail Price: $14.99**

Montana Mining Books

<u>A History of Butte Montana: The World's Greatest Mining Camp</u> - First published in 1900 by H.C. Freeman, this important publication sheds a bright light on one of the most important mining areas in the history of The West. Together with his insights, as well as rare photographs of the periods, Harry Freeman describes Butte and its vicinity from its early beginnings, right up to its flush years when copper flowed from its mines like a river. At the time of publication, Butte, Montana was known worldwide as "The Richest Mining Spot On Earth" and produced not only vast amounts of copper, but also silver, gold and other metals from its mines. Freeman illustrates, with great detail, the most important mines in the vicinity of Butte, providing rare details on their owners, their history and most importantly, how the mines operated and how their treasures were extracted. Of particular interest are the dozens of rare photographs that depict mines such as the famous Anaconda, the Silver Bow, the Smoke House, Moose, Paulin, Buffalo, Little Minah, the Mountain Consolidated, West Greyrock, Cora, the Green Mountain, Diamond, Bell, Parnell, the Neversweat, Nipper, Original and many others.   **8.5" X 11", 142 ppgs. Retail Price: $12.99**

<u>The Butte Mining District of Montana</u> - This important publication on Montana Mining has not been available for over a century. Included are rare insights into the gold, copper and silver mines of Butte, Montana together with hard to find maps and photographs. Some of the topics include the early history of gold, silver and copper mining in the Butte area, insight into the geology of its mining areas, the local distribution of gold, silver and copper ores, as well their composition and how to identify them. Also included are detailed facts about the mines in the Butte Mining District, including the famous Anaconda Mine, Gagnon, Parrot, Blue Vein, Moscow, Poulin, Stella, Buffalo, Green Mountain, Wake Up Jim, the Diamond-Bell Group, Mountain Consolidated, East Greyrock, West Greyrock, Snowball, Corra, Speculator, Adirondack, Miners Union, the Jessie-Edith May Group, Otisco, Iduna, Colorado, Lizzie, Cambers, Anderson, Hesperus, Preferencia and dozens of others. **8.5" X 11", 298 ppgs. Retail Price: $24.99**

<u>Mines of the Helena Mining Region of Montana</u> - This important publication on Montana Mining has not been available for over a century. Included are rare insights into the gold, copper and silver mines of the vicinity of Helena, Montana, including the Marysville Mining District, Elliston Mining District, Rimini Mining District, Helena Mining District, Clancy Mining District, Wickes Mining District, Boulder and Basin Mining Districts and the Elkhorn Mining District. Some of the topics include the early history of gold, silver and copper mining in the Helena area, insight into the geology of its mining areas, the local distribution of gold, silver and copper ores, as well their composition and how to identify them. Also included are detailed facts, history, geology and locations of over one hundred gold, silver and copper mines in the area . **8.5" X 11", 162 ppgs, Retail Price: $14.99**

<u>Mines and Geology of the Garnet Range of Montana</u> - This important publication on Montana Mining has not been available for over a century. Included are rare insights into the gold, copper and silver mines of the vicinity of this important mining area of Montana. Some of the topics include the early history of gold, silver and copper mining in the Garnet Mountains, insight into the geology of its mining areas, the local distribution of gold, silver and copper ores, as well their composition and how to identify them. Also included are detailed facts, history, geology and locations of numerous gold, silver and copper mines in the area . **8.5" X 11", 100 ppgs, Retail Price: $11.99**

<u>Mines and Geology of the Philipsburg Quadrangle of Montana</u> - This important publication on Montana Mining has not been available for over a century. Included are rare insights into the gold, copper and silver mines of the vicinity of this important mining area of Montana. Some of the topics include the early history of gold, silver and copper mining in the Philipsburg Quadrangle, insight into the geology of its mining areas, the local distribution of gold, silver and copper ores, as well their composition and how to identify them. Also included are detailed facts, history, geology and locations of over one hundred gold, silver and copper mines in the area **8.5" X 11", 290 ppgs, Retail Price: $24.99**

<u>Geology of the Marysville Mining District of Montana</u> - Included are rare insights into the mining geology of the Marysville Mining District. Some of the topics include the early history of gold, silver and copper mining in the area, insight into the geology of its mining areas, the local distribution of gold, silver and copper ores, as well their composition and how to identify them. Also included are detailed facts, history, geology and locations of gold, silver and copper mines in the area **8.5" X 11", 198 ppgs, Retail Price: $19.99**

<u>**The Geology and Mines of Northern Idaho and North Western Montana**</u>

See listing under Idaho.

Nevada Mining Books

<u>**The Bull Frog Mining District of Nevada**</u> - Unavailable since 1910, this publication was originally compiled by the United States Department of Interior. This volume also includes important insights into the geologic formations, faults and other aspects of economic geology in this Nevada mining district. Of particular interest are the fine details on many mines in the area, including their locations, histories, development and mineralization. Some of the mines featured include the National Bank Mine, Providence, Gibraltor, Tramps, Denver, Original Bullfrog, Gold Bar, Mayflower, Homestake-King and other mines and prospects. **8.5" X 11", 152 ppgs, Retail Price: $14.99**

<u>History of the Comstock Lode</u> - Unavailable since 1876, this publication was originally released by John Wiley & Sons. This volume also includes important insights into the famous Comstock Lode of Nevada that represented the first major silver discovery in the United States. During its spectacular run, the Comstock produced over 192 million ounces of silver and 8.2 million ounces of gold. Not only did the Comstock result in one of the largest mining rushes in history and yield immense fortunes for its owners, but it made important contributions to the development of the State of Nevada, as well as neighboring California. Included here are important details on not only the early development and history of the Comstock, but also rare early insight into its mines, ore and its geology.**8.5" X 11", 244 ppgs, Retail Price: $19.99**

Colorado Mining Books

<u>**Ores of The Leadville Mining District**</u> - Unavailable since 1926, this publication was originally compiled by the United States Department of Interior. This volume also includes important insights into the ores and mineralization of the Leadville Mining District in Colorado. Topics include historic ore prospecting methods, local geology, insights into ore veins and stockworks, the local trend and distribution of ore channels, reverse faults, shattered rock above replacement ore bodies, mineral enrichment in oxidized and sulphide zones and more. **8.5" X 11", 66 ppgs, Retail Price: $8.99**

<u>**Mining in Colorado**</u> - Unavailable since 1926, this publication was originally compiled by the United States Department of Interior. This volume also includes important insights into the mining history of Colorado from its early beginnings in the 1850's right up to the mid 1920's. Not only is Colorado's gold mining heritage included, but also its silver, copper, lead and zinc mining industry. Each mining area is treated separately, detailing the development of Colorado's mines on a county by county basis. **8.5" X 11", 284 ppgs, Retail Price: $19.99**

<u>Gold Mining in Gilpin County Colorado</u> - Unavailable since 1876, this publication was originally compiled by the Register Steam Printing House of Central City, Colorado. A rare glimpse at the gold mining history and early mines of Gilpin County, Colorado from their first discovery in the 1850's up to the "flush years" of the mid 1870's. Of particular interest is the history of the discovery of gold in Gilpin County and details about the men who made those first strikes. Special focus is given to the early gold mines and first mining districts of the area, many of which are not detailed in other books on Colorado's gold mining history. **8.5" X 11", 156 ppgs, Retail Price: $12.99**

<u>Mining in the Gold Brick Mining District of Colorado</u> - Important insights into the history of the Gold Brick Mining District, as well as its local geography and economic geology. Also included are the histories and locations of historic mines in this important Colorado Mining District, including the Cortland, Carter, Raymond, Gold Links, Sacramento, Bassick, Sandy Hook, Chronicle, Grand Prize, Chloride, Granite Mountain, Lucille, Gray Mountain, Hilltop, Maggie Mitchell, Silver Islet, Revenue, Roosevelt, Carbonate King and others. In addition to hardrock mining, are also included are details on gold placer mining in this portion of Colorado. **8.5" X 11", 140 ppgs, Retail Price: $12.99**

Washington Mining Books

<u>**The Republic Mining District of Washington**</u> - Unavailable since 1910, this important publication was originally published by the Washington Geologic Survey and has been unavailable for a century. Topics include the geology, rock formations and the formation of ore deposits in this important mining area of Washington State. Also included are hard to find details on the geology, history and locations of dozens of mines in the area. Some of the mines featured include the New Republic Mine, Ben Hur, Morning Glory, the South Republic Mine, Quilp, Surprise, Black Tail, Lone Pine, San Poil, Mountain Lion, Tom Thumb, Elcaliph and many others. **8.5" X 11", 94 ppgs, Retail Price: $10.99**

The Myers Creek and Nighthawk Mining Districts of Washington - Unavailable since 1911, this important publication was originally published by the Washington Geologic Survey and has been unavailable for a century. Topics include the geology, rock formations and the formation of ore deposits in these important mining areas of Washington State. Also included are hard to find details on the geology, history and locations of dozens of mines in the area. Some of the mines featured include the Grant Mine, Monterey, Nip and Tuck, Myers Creek, Number Nine, Neutral, Rainbow, Aztec, Crystal Butte, Apex, Butcher Boy, Molson, Mad River, Olentangy, Delate, Kelsey, Golden Chariot, Okanogan, Ohio, Forty-Ninth Parallel, Nighthawk, Favorite, Little Chopaka, Summit, Number One, California, Peerless, Caaba, Prize Group, Ruby, Mountain Sheep, Golden Zone, Rich Bar, Similkameen, Kimberly, Triune, Hiawatha, Trinity, Hornsilver, Maquae, Bellevue, Bullfrog, Palmer Lake, Ivanhoe, Copper World and many others.
 8.5" X 11", 136 ppgs, Retail Price: $12.99

The Blewett Mining District of Washington - Unavailable since 1911, this important publication was originally published by the Washington Geologic Survey and has been unavailable for a century. Topics include the geology, rock formations and the formation of ore deposits in this important mining area of Washington State. Also included are hard to find details on the geology, history and locations of dozens of mines in the area. Some of the mines featured include the Washington Meteor, Alta Vista, Pole Pick, Blinn, North Star, Golden Eagle, Tip Top, Wilder, Golden Guinea, Lucky Queen, Blue Bell, Prospect, Homestake, Lone Rock, Johnson, and others. **8.5" X 11", 134 ppgs, Retail Price: $12.99**

Silver Mining In Washington - Unavailable since 1955, this important publication was originally published by the Washington Geologic Survey. Featured are the hard to find locations and details pertaining to Washington's silver mines. **8.5" X 11", 180 ppgs, Retail Price: $15.99**

The Mines of Snohomish County Washington - Unavailable since 1942, this important publication was originally published by the Washington Geologic Survey and has been unavailable for seventy years. Featured are details on a large number of gold, silver, copper, lead and other metallic mineral mines. Included are the locations of each historic mine, along with information on the commodity produced. **8.5" X 11", 98 ppgs, Retail Price: $10.99**

The Mines of Chelan County Washington - Unavailable since 1943, this important publication was originally published by the Washington Geologic Survey and has been unavailable for seventy years. Featured are details on a large number of gold, silver, copper, lead and other metallic mineral mines. Included are the locations of each historic mine, along with information on the commodity. **8.5" X 11", 88 ppgs, Retail Price: $9.99**

Metal Mines of Washington - Unavailable since 1921, this important publication was originally published by the Washington Geologic Survey and has been unavailable for nearly ninety years. Widely considered a masterpiece on the Washington Mining Industry, "Metal Mines of Washington" sheds light on the important details of Washington's early mining years. Featured are details on hundreds of gold, silver, copper, lead and other metallic mineral mines. Included are hard to find details on the mineral resources of this state, as well as the locations of historic mines. Lavishly illustrated with maps and historic photos and complete with a glossary to explain any technical terms found in the text, this is one of the most important works on mining in the State of Washington. No prospector or miner should be without it if they are interested in mining in Washington. **8.5" X 11", 396 ppgs, Retail Price: $24.99**

Gem Stones In Washington - Unavailable since 1949, this important publication was originally published by the Washington Geologic Survey and has been unavailable since first published. Included are details on where to find naturally occurring gem stones in the State of Washington, including quartz crystal, amethyst, smoky quartz, milky quartz, agates, bloodstone, carnelian, chert, flint, jasper, onyx, petrified wood, opal, fire opal, hyalite and others. **8.5" X 11", 54 ppgs, Retail Price: $8.99**

The Covada Mining District of Washington - Unavailable since 1913, this important publication was originally published by the Washington Geologic Survey and has been unavailable for a century. Topics include the geology, rock formations and the formation of ore deposits in this important mining area of Washington State. Also included are hard to find details on the geology, history and locations of dozens of mines in the area. Some of the mines featured include the Admiral, Advance, Algonkian, Big Bug, Big Chief, Big Joker, Black Hawk, Black Tail, Black Thorn, Captain, Cherokee Strip, Colorado, Dan Patch, Dead Shot, Etta, Good Ore, Greasy Run, Great Scott, Idora, IXL, Jay Bird, Kentucky Bell, King Solomon, Laurel, Laura S, Little Jay, Meteor, Neglected, Northern Light, Old Nell, Plymouth Rock, Polaris, Quandary, Reserve, Shoo Fly, Silver Plume, Three Pines, Vernie, White Rose and dozens of others. **8.5" X 11", 114 ppgs, Retail Price: $10.99**

The Index Mining District of Washington - Unavailable since 1912, this important publication was originally published by the Washington Geologic Survey and has been unavailable for a century. Topics include the geology, rock formations and the formation of ore deposits in this important mining area of Washington State. Also included are hard to find details on the geology, history and locations of dozens of mines in the area. Some of the mines featured include the Sunset, Non-Pareil, Ethel Consolidated, Kittaning, Merchant, Homestead, Co-operative, Lost Creek, Uncle Sam, Calumet, Florence-Rae, Bitter Creek, Index Peacock, Gunn Peak, Helena, North Star, Buckeye. Copper Bell, Red Cross and others. **8.5" X 11", 114 ppgs, Retail Price: $11.99**

Mining & Mineral Resources of Stevens County Washington - Unavailable since 1920, this important publication was originally published by the Washington Geologic Survey and has been unavailable for a century. Topics include the geology, rock formations and the formation of ore deposits in these important mining areas of Washington State. Also included are hard to find details on the geology, history and locations of hundreds of mines in the area. 8.5" X 11", 372 ppgs, Retail Price: $24.99

The Mines and Geology of the Loomis Quadrangle Okanogan County, Washington - Unavailable since 1972, this important publication was originally published by the Washington Geologic Survey and has been unavailable for a century. Topics include the geology, rock formations and the formation of ore deposits in this important mining area of Washington State. Also included are hard to find details on the geology, history and locations of dozens of gold, copper, silver and other mines in the area. 8.5" X 11", 150 ppgs, Retail Price: $12.99

The Conconully Mining District of Okanogan County Washington - Unavailable since 1973, this important publication was originally published by the Washington Geologic Survey and has been unavailable for a century. Topics include the geology, rock formations and the formation of ore deposits in this important mining area of Washington State, which also includes Salmon Creek, Blue Lake and Galena. Also included are hard to find details on the geology, mining history and locations of dozens of mines in the area. Some of the mines include Arlington, Fourth of July, Sonny Boy, First Thought, Last Chance, War Eagle-Peacock, Wheeler, Mohawk, Lone Star, Woo Loo Moo Loo, Keystone, Hughes, Plant-Callahan, Johnny Boy, Leuena, Gubser, John Arthur, Tough Nut, Homestake, Key and many others 8.5" X 11", 68 ppgs, Retail Price: $8.99

Wyoming Mining Books

Mining in the Laramie Basin of Wyoming - Unavailable since 1909, this publication was originally compiled by the United States Department of Interior. Also included are insights into the mineralization and other characteristics of this important mining region, especially in regards to coal, limestone, gypsum, bentonite clay, cement, sand, clay and copper. 8.5" X 11", 104 ppgs, Retail Price: $11.99

New Mexico Mining Books

The Mogollon Mining District of New Mexico - Unavailable since 1927, this important publication was originally published by the US Department of Interior and has been unavailable for 80 years. Topics include the geology, rock formations and the formation of ore deposits in this important mining area in New Mexico. Of particular focus is information on the history and production of the ore deposits in this area, their form and structure, vein filling, their paragenesis, origins and ore shoots, as well as oxidation and supergene enrichment. Also included are hard to find details, including the descriptions and locations of numerous gold, silver and other types of mines, including the Eureka, Pacific, South Alpine, Great Western, Enterprise, Buffalo, Mountain View, Floride, Gold Dust, Last Chance, Deadwood, Confidence, Maud S., Deep Down, Little Fanney, Trilby, Johnson, Alberta, Comet, Golden Eagle, Cooney, Queen, the Iron Crown, Eberle, Clifton, Andrew Jackson mine, Mascot and others. 8.5" X 11", 144 ppgs, Retail Price: $12.99

The Percha Mining District of Kingston New Mexico - Unavailable since 1883, this important publication was originally published by the Kingston Tribune and has been unavailable for over one hundred and thirty five years. Having been written during the earliest years of gold and silver mining in the Percha Mining District, unlike other books on the subject, this work offers the unique perspective of having actually been written while the early mining history of this area was still being made. In fact, the work was written so early in the development of this area that many of the notable mines in the Percha District were less than a few years old and were still being operated by their original discoverers with the same enthusiasm as when they were first located. Included are hard to find details on the very earliest gold and silver mines of this important mining district near Kingston in Sierra County, New Mexico. 8.5" X 11", 68 ppgs, Retail Price: $9.99

East Coast Mining Books

The Gold Fields of the Southern Appalachians - Unavailable since 1895, this important publication was originally published by the US Department of Interior and has been unavailable for nearly 120 years. Topics include the geology, rock formations and the formation of ore deposits in this important mining area of the American South. Of particular focus is information on the history and statistics of the ore deposits in this area, their form and structure and veins. Also included are details on the placer gold deposits of the region. The gold fields of the Georgian Belt, Carolinian Belt and the South Mountain Mining District of North Carolina are all treated in descriptive detail. Included are hard to find details, including the descriptions and locations of numerous gold mines in Georgia, North Carolina and elsewhere in the American South. Also included are details on the gold belts of the British Maritime Provinces and the Green Mountains. 8.5" X 11", 104 ppgs, Retail Price: $9.99

Gold Rush Tales Series

Millions in Siskiyou County Gold - In this first volume of the "Gold Rush Tales" series, leading mining historian and editor Kerby Jackson, introduces us to the story of how millions of dollars worth of gold was discovered in Siskiyou County during the California Gold Rush. Lavishly illustrated with photos from the 19th Century, this hard to find information was first published in 1897 and sheds important light onto the gold rush era in Siskiyou County, California and the experiences of the men who dug for the gold and actually found it. **8.5" X 11", 82 ppgs, Retail Price: $9.99**

The California Rand in the Days of '49 - In this second volume of the "Gold Rush Tales" series, leading mining historian and editor Kerby Jackson, introduces us to four tales from the California Gold Rush. Lavishly illustrated with photos from the 19th Century, this hard to find information was first published in 1890's and includes the stories of "California's Rand", details about Chinese miners, how one early miner named Baker struck it rich and also the story of Alphonzo Bowers, who invented the first hydraulic gold dredge. **8.5" X 11", 54 ppgs, Retail Price: $9.99**

More Mining Books

Prospecting and Developing A Small Mine - Topics covered include the classification of varying ores, how to take a proper ore sample, the proper reduction of ore samples, alluvial sampling, how to understand geology as it is applied to prospecting and mining, prospecting procedures, methods of ore treatment, the application of drilling and blasting in a small mine and other topics that the small scale miner will find of benefit. **8.5" X 11", 112 ppgs, Retail Price: $11.99**

Timbering For Small Underground Mines - Topics covered include the selection of caps and posts, the treatment of mine timbers, how to install mine timbers, repairing damaged timbers, use of drift supports, headboards, squeeze sets, ore chute construction, mine cribbing, square set timbering methods, the use of steel and concrete sets and other topics that the small underground miner will find of benefit. This volume also includes twenty eight illustrations depicting the proper construction of mine timbering and support systems that greatly enhance the practical usability of the information contained in this small book. **8.5" X 11", 88 ppgs. Retail Price: $10.99**

Timbering and Mining - A classic mining publication on Hard Rock Mining by W.H. Storms. Unavailable since 1909, this rare publication provides an in depth look at American methods of underground mine timbering and mining methods. Topics include the selection and preservation of mine timbers, drifting and drift sets, driving in running ground, structural steel in mine workings, timbering drifts in gravel mines, timbering methods for driving shafts, positioning drill holes in shafts, timbering stations at shafts, drainage, mining large ore bodies by means of open cuts or by the "Glory Hole" system, stoping out ore in flat or low lying veins, use of the "Caving System", stoping in swelling ground, how to stope out large ore bodies, Square Set timbering on the Comstock and its modifications by California miners, the construction of ore chutes, stoping ore bodies by use of the "Block System", how to work dangerous ground, information on the "Delprat System" of stoping without mine timbers, construction and use of headframes and much more. This volume provides a reference into not only practical methods of mining and timbering that may be employed in narrow vein mining by small miners today, but also rare insights into how mines were being worked at the turn of the 19th Century. **8.5" X 11", 288 ppgs. Retail Price: $24.99**

A Study of Ore Deposits For The Practical Miner - Mining historian Kerby Jackson introduces us to a classic mining publication on ore deposits by J.P. Wallace. First published in 1908, it has been unavailable for over a century. Included are important insights into the properties of minerals and their identification, on the occurrence and origin of gold, on gold alloys, insights into gold bearing sulfides such as pyrites and arsenopyrites, on gold bearing vanadium, gold and silver tellurides, lead and mercury tellurides, on silver ores, platinum and iridium, mercury ores, copper ores, lead ores, zinc ores, iron ores, chromium ores, manganese ores, nickel ores, tin ores, tungsten ores and others. Also included are facts regarding rock forming minerals, their composition and occurrences, on igneous, sedimentary, metamorphic and intrusive rocks, as well as how they are geologically disturbed by dikes, flows and faults, as well as the effects of these geologic actions and why they are important to the miner. Written specifically with the common miner and prospector in mind, the book will help to unlock the earth's hidden wealth for you and is written in a simple and concise language that anyone can understand. **8.5" X 11", 366 ppgs. Retail Price: $24.99**

Mine Drainage - Unavailable since 1896, this rare publication provides an in depth look at American methods of underground mine drainage and mining pump systems. This volume provides a reference into not only practical methods of mining drainage that may be employed in narrow vein mining by small miners today, but also rare insights into how mines were being worked at the turn of the 19th Century. **8.5" X 11", 218 ppgs. Retail Price: $24.99**

Fire Assaying Gold, Silver and Lead Ores - Unavailable since 1907, this important publication was originally published by the Mining and Scientific Press and was designed to introduce miners and prospectors of gold, silver and lead to the art of fire assaying. Topics include the fire assaying of ores and products containing gold, silver and lead; the sampling and preparation of ore for an assay; care of the assay office, assay furnaces; crucibles and scorifiers; assay balances; metallic ores; scorification assays; cupelling; parting' crucible assays, the roasting of ores and more. This classic provides a time honored method of assaying put forward in a clear, concise and easy to understand language that will make it a benefit to even beginners. **8.5″ X 11″, 96 ppgs. Retail Price: $11.99**

Methods of Mine Timbering - Originally published in 1896, this important publication on mining engineering has not been available for nearly a century. Included are rare insights into historical methods of timbering structural support that were used in underground metal mines during the California that still have a practical application for the small scale hardrock miner of today. **8.5″ X 11″, 94 ppgs. Retail Price: $10.99**

The Enrichment of Copper Sulfide Ores - First published in 1913, it has been unavailable for over a century. Topics include the definition and types of ore enrichment, the oxidation of copper ores, the precipitation of metallic sulfides. Also included are the results of dozens of lab experiments pertaining to the enrichment of sulfide ores that will be of interest to the practical hard rock mine operator in his efforts to release the metallic bounty from his mine's ore. **8.5″ X 11″, 92 ppgs. Retail Price: $9.99**

A Study of Magmatic Sulfide Ores - Unavailable since 1914, this rare publication provides an in depth look at magmatic sulfide ores. Some of the topics included are the definition and classification of magmatic ores, descriptions of some magmatic sulfide ore deposits known at the time of publication including copper and nickel bearing pyrrohitic ore bodies, chalcopyrite-bornite deposits, pyritic deposits, magnetite-ileminite deposits, chromite deposits and magmatic iron ore deposits. Also included are details on how to recognize these types of ore deposits while prospecting for valuable hardrock minerals. **8.5″ X 11″, 138 ppgs. Retail Price: $11.99**

The Cyanide Process of Gold Recovery - Unavailable since 1894 and released under the name "The Cyanide Process: Its Practical Application and Economical Results", this rare publication provides an in depth look at the early use of cyanide leaching for gold recovery from hardrock mine ores. This volume provides a reference into the early development and use of cyanide leaching to recover gold. **8.5″ X 11″, 162 ppgs. Retail Price: $14.99**

California Gold Milling Practices - Unavailable since 1895 and released under the name "California Gold Practices", this rare publication provides an in depth look at early methods of milling used to reduce gold ores in California during the late 19th century. This volume provides a reference into the early development and use of milling equipment during the earliest years of the California Gold Rush up to the age of the Industrial Revolution. Much of the information still applies today and will be of use to small scale miners engaging in hardrock mining. **8.5″ X 11″, 104 ppgs. Retail Price: $10.99**